CUNARD

THE MOST FAMOUS OCEAN LINERS IN THE WORLD ™

Reference Library

FOR REFERENCE ONLY

PLEASE DO NOT REMOVE THIS BOOK
FROM THE LIBRARY

Library by Ocean Books
www.oceanbooks.com

Applied Oceanography

APPLIED OCEANOGRAPHY

JOSEPH M. BISHOP

Federal Emergency Management Agency
and
Department of Civil Engineering
Catholic University of America

A WILEY-INTERSCIENCE PUBLICATION

JOHN WILEY & SONS

New York • Chichester • Brisbane • Toronto • Singapore

Library of Congress Cataloging in Publication Data:

Bishop, Joseph M. (Joseph Michael), 1943–
  Applied oceanography.

  (Ocean engineering: a Wiley series)
  Includes bibliographical references and index.
  1. Oceanography. 2. Marine resources. I. Title. II. Series.

GC1015.2.B57 1984   551.46   83-26091
ISBN 0-471-87445-0

Printed in the United States of America

10 9 8 7 6 5 4 3 2 1

SERIES PREFACE

Ocean engineering is both old and new. It is old in that man has concerned himself with specific problems in the ocean for thousands of years. Ship building, prevention of beach erosion, and construction of offshore structures are just a few of the specialties that have been developed by engineers over the ages. Until recently, however, these efforts tended to be restricted to specific areas. Within the past decade an attempt has been made to coordinate the activities of all technologists in ocean work, calling the entire field "ocean engineering." Here we have its newness.

Ocean Engineering: A Wiley Series has been created to introduce engineers and scientists to the various areas of ocean engineering. Books in this series are so written as to enable engineers and scientists easily to learn the fundamental principles and techniques of a specialty other than their own. The books can also serve as textbooks in advanced undergraduate and introductory graduate courses. The topics to be covered in this series include ocean engineering wave mechanics, marine corrosion, coastal engineering, dynamics of marine vehicles, offshore structures, and geotechnical or seafloor engineering. We think that this series fills a great need in the literature of ocean technology.

MICHAEL E. McCORMICK, EDITOR
RAMESWAR BHATTACHARYYA, ASSOCIATE EDITOR

November 1972

PREFACE

Oceanography books fall into two main categories. Those of the first type detail scientific relationships, giving little attention to their applications; those of the second type develop specific applications, usually with little or no concern for the scientific aspects of the topic. Readers of the first category often find it difficult to relate the material presented to real-world problems. Readers of the second category may never fully appreciate the scientific studies that have laid the groundwork for a particular application.

Given this perspective, the main objectives of this book are:

1. To provide the ocean-user community with a reference work that relates applied oceanography to the science of physical oceanography, where applied oceanography is defined as a system with physical oceanography, marine ecology, economics, and government policy as its four components.

2. To provide an introductory text suitable for both the student and the practitioner of physical oceanography who may also have an interest in the ultimate application of this knowledge to contemporary problems in marine pollution, marine resources, and marine transportation.

3. To separate background scientific information from its applications, giving adequate attention to each.

4. To allow those presently engaged in some aspect of applied oceanography to develop a basic understanding of additional topics, and thereby to promote "cross-fertilization" within the field. For example, the major naval powers of the world have developed advanced techniques to predict ocean thermal structure for anti-submarine operations. The same prediction techniques could also be applied to locating temperature-sensitive fish species.

The book is organized into two major parts. In Part One, consisting of five chapters, basic concepts of physical oceanography are outlined, with emphasis on the development of information related to the applied topics covered in later chapters. In Part Two, consisting of nine chapters, applied oceanography is addressed, with examples of applications in marine pollution, marine resources, and marine transportation.

The information presented in the following pages represents an effort to communicate my understanding of the newly developing field of applied oceanography. It does not represent a complete and comprehensive development of individual topics within the field, but rather a summary of information considered appropriate based on my background and interests. Regardless, the topics as developed will serve as a point of departure for those interested in, working in, or in need of information on how physical oceanography can be applied to solve real-world problems of contemporary interest.

JOSEPH M. BISHOP

Washington, D.C.
April 1984

ACKNOWLEDGMENTS

Throughout the preparation of this book I tried to condense the work of a number of investigators engaged in various aspects of applied oceanography. I acknowledge their contributions and hope that I have reported their work adequately and correctly. I was also fortunate to be able to draw upon the fine efforts of a few special people who worked with me to get the manuscript "clean" enough to send to the typesetter. The first, Elizabeth Haynes, of the National Oceanographic and Atmospheric Administration, is a meteorologist turned fisheries oceanographer. Elizabeth carefully used her sharp red pencil on at least two drafts of the original text. I thank her for giving generously of her time to review the manuscript. Second, I would like to thank Kathryn Bush, a physical oceanographer with Marine Environments Corporation, for her careful technical editing of the final manuscript. In addition, credit must go to Amikam Gilad of Marine Environments Corporation, who completed many of the illustrations in the book. Thanks also go to Dr. Richard James of the U.S. Navy, Dr. Carvel Blair of Old Dominion University, and Mr. Richard Hayes of the U.S. Coast Guard for reviewing portions of the manuscript. Finally, I would like to thank Dr. Marshall Earle, President of Marine Environments Corporation, for his review of the technical aspects of the book.

J.M.B.

CONTENTS

Applied Oceanography

A SURVEY OF PHYSICAL OCEANOGRAPHY

A main objective of the book is to introduce the field of applied oceanography to the ocean-user community. The subject is addressed with a primary focus on physical oceanography. Therefore, a survey of physical oceanography is presented to give a basis from which to develop various topics in applied oceanography. This information is intended to serve as a review for those involved in physical oceanography and as introductory material for those not well versed in the subject. The major areas covered include basic meteorology and oceanography, ocean currents, ocean waves, and the advection and mixing of contaminants at sea. The survey will be developed with an applications orientation, thereby giving a different perspective from that usually presented in typical reference works on physical oceanography.

THE OCEANS:
SEA AND AIR

In this chapter we will develop fundamental relationships important to both meteorology and oceanography. We begin with a discussion of how the sun provides the radiant energy that ultimately drives the circulation of the oceans of sea and air that cover the globe. Accordingly, atmospheric winds and oceanic currents are described as nature's way of spreading the excess low-latitude solar energy across the Earth's surface.

Meteorologists and oceanographers have classically employed specialized charts of such physical parameters as wind, current, temperature, pressure, and density to study oceanic and atmospheric exchange processes. For example, sea–air temperature difference charts are used to estimate heat exchange across the sea–air interface, and surface wind charts are used to calculate momentum flux to ocean waves and currents. This approach has led to the application of relationships commonly used in meterology, a more advanced environmental science, to physical oceanography. Analogies between air masses and water masses and between atmospheric fronts and oceanic fronts have provided valuable insights into the physics and dynamics of the ocean.

Meteorological parameters commonly charted include air temperature, air pressure, and water-vapor content, which are related to air density by a theoretically derived equation of state. Distributions of these variables of state are dynamically related to winds. Oceanographic parameters charted include sea temperature, sea pressure, and salinity, which are related to sea water density by an empirical equation of state. Distributions of the oceanic variables of state are dynamically related to ocean currents.

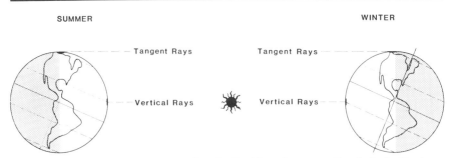

Figure 1.1. Radiation from the sun strikes the Earth's surface more directly in the Northern Hemisphere in the summer than in the winter.

1.1.   THE ENERGY SOURCE

The Earth travels in an elliptical orbit around the sun at an average distance of about 150 million kilometers. It also rotates on its axis once each day, producing alternate days and nights. The Earth's axis of rotation is inclined 23.5° to the plane of its orbit around the sun. The annual orbit, coupled with the inclination of the Earth's axis, produces seasonal variations in the intensity of incoming solar radiation, as shown in Figure 1.1.

The sun radiates electromagnetic energy over a spectrum of wavelengths from short-wave, high-frequency x-rays to long-wave, lower-frequency infrared waves and even longer radio waves. Visible light, a mixture of such familiar colors as violet, blue, green, and red, is located in the highest energy portion of this spectrum. Approximately 40 percent of the total solar radiation received at the Earth's upper atmosphere is either scattered by atmospheric particles or reflected back into space. In addition, about 15 percent of the incoming solar radiation is absorbed by the atmosphere. The remaining energy reaches and heats the Earth's surface. The Earth then re-radiates this heat in the infrared portion of the electromagnetic spectrum. Outgoing terrestrial radiation is selectively absorbed by atmospheric carbon dioxide and water vapor in the lower atmosphere. This process is commonly referred to as the "greenhouse effect," since the atmosphere acts like the glass of a greenhouse, admitting incoming and trapping outgoing radiation. The greenhouse effect accounts for the typical decrease in temperature with altitude in the Earth's lower atmosphere.

1.2.   THE GENERAL CIRCULATION OF THE ATMOSPHERE

The Earth's equatorial regions are far warmer than the poles because they receive direct solar radiation, as illustrated in Figure 1.1. At the Earth's surface in equatorial latitudes vast amounts of warmed air rise into the upper atmosphere, forming a surface low-pressure region known as the Intertropical Convergence Zone (ITCZ). The ITCZ is generally located north of the geographic equator, and is often visible on weather satellite images as a band of clouds that nearly circles the globe. The rising air reaches altitudes of 15 to 20 km and then flows north and south toward the poles. A portion of this air sinks back toward the Earth's surface at approximately 30°N and 30°S latitudes, producing the subtropical high-pressure zones. The cold dense air over the polar regions also sinks, producing additional surface high-pressure zones. Subtropical high-pressure zones that are formed over warm ocean areas produce warm, moist air masses, while high-pressure zones formed over the poles generally produce cold, dry air masses. The subtropical and polar high-pressure regions are separated by bands of low pressure located at about 60°N and 60°S latitudes. These regions, known as the polar fronts, tend to move poleward in the summer and equatorward in the winter. Like the ITCZ, the polar fronts can be identified by their cloud cover, storminess, upward air flow at the Earth's surface, and relatively low surface atmospheric pressure.

Pressure differences between these global high- and low-pressure zones produce a wind flowing toward lower pressure. The greater the pressure gradient (the pressure change over a given distance), the faster this wind blows. The Earth's west to east rotation modifies the direct flow toward low pressure. Viewed from the Earth, there is a turning of the original wind direction to the right of the flow in the Northern Hemisphere and to the left in the Southern Hemisphere. This deflection is known as the Coriolis effect, and plays a key role in the dynamics of large-scale atmospheric and oceanic flow. In the Northern Hemisphere the Coriolis effect causes poleward air flow to turn east and equatorward air flow to turn west. The opposite occurs in the Southern Hemisphere.

As an example of how global-scale pressure gradients and the Coriolis effect combine to produce global-scale wind patterns, consider the northeast trade winds in the Northern Hemisphere. Between the subtropical high-pressure zone (at 30°N) and the ITCZ the pressure gradient is directed toward the equator. The wind initially blows in that direction, but is deflected to the right (west) by the Coriolis effect, resulting in the northeast trade-wind belt. The prevailing westerlies and the polar easterlies of both hemispheres can be explained similarly. In short, the general circulation of the atmosphere, as shown in Figure 1.2, can be described in simple terms as a response to global-scale pressure gradients and the Earth's rotation.

1.3. UNITS AND DIMENSIONS

The CGS system of units, which consists of the metric units centimeters (cm), grams (g), and seconds (s), is commonly used in meteorology and oceanography and will be used in this book. In addition, the MKS system, consisting of the meter (1 m = 100 cm), the kilogram (1 kg = 1000 g), and the

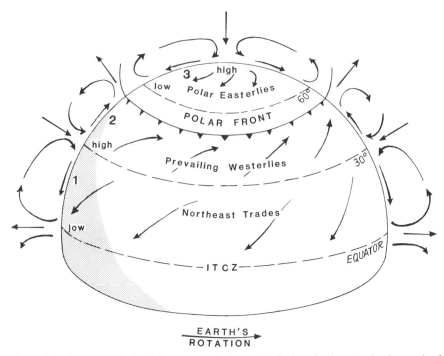

Figure 1.2.   The general circulation of the Northern Hemisphere is shown to be the result of the three global-scale pressure gradients (numbered 1, 2, and 3) and the Earth's rotation.

second, will be used interchangeably with the CGS system, with the appropriate conversion. Units such as the kilometer (1 km = 1000 m), foot (30.48 cm), nautical mile (1853m = 1.853 km), statute mile (1609 m = 1.609 km), and knot (51.47 cm/s) may also be used on occasion.

Quantities can be expressed in terms of the basic dimensions of length (L), mass (M), time (T), and temperature (t). Some of the more commonly used metric quantities are shown in Table 1.1. Attention to the units and dimensions of a quantity can help in the verification of equations and in some cases in the development of new relationships among variables.

1.4.   THE VARIABLES OF STATE

A classical approach to oceanography and meterology is to study fluid processes by mapping the variables of state. Observed data are generally plotted on synoptic charts, which show observations taken at about the same time, or on climatic charts, which show observations that have been averaged over time and space. These charts have been used to develop mathematical relations between the variables of state and to gain insight into fluid-dynamical processes. For example, the average January sea- and air-

TABLE 1.1.   DIMENSIONS AND METRIC UNITS OF COMMONLY USED QUANTITIES IN METEOROLOGY AND OCEANOGRAPHY

Quantity	Dimension	CGS Unit	MKS Unit
Length	L	cm	m
Mass	M	g	kg
Time	T	s	s
Temperature	t	a	a
Volume	L^3	cm^3	m^3
Density	$\dfrac{M}{L^3}$	$\dfrac{g}{cm^3}$	$\dfrac{kg}{m^3}$
Specific volume	$\dfrac{L^3}{M}$	$\dfrac{cm^3}{g}$	$\dfrac{m^3}{kg}$
Velocity	$\dfrac{L}{T}$	$\dfrac{cm}{s}$	$\dfrac{m}{s}$
Acceleration	$\dfrac{L}{T^2}$	$\dfrac{cm}{s^2}$	$\dfrac{m}{s^2}$
Force	$\dfrac{ML}{T^2}$	$\dfrac{g\ cm}{s^2}$ (dyne)	$\dfrac{kg\ m}{s^2}$ (newton)
Pressure	$\dfrac{M}{LT^2}$	$\dfrac{g}{cm\ s^2}\left(\dfrac{dyne}{cm^2}\right)$	$\dfrac{kg}{ms^2}$ (pascal)
Work	$\dfrac{ML^2}{T^2}$	$\dfrac{g\ cm^2}{s^2}$ (erg)	$\dfrac{kgm^2}{s^2}$ (joule)

aTemperature may be expressed in the Celsius (°C) or Fahrenheit (°F) scales. Conversion from one to another scale is done using

$$\frac{°C}{100°} = \frac{°F - 32°}{180°}$$

surface temperature patterns are shown in Figure 1.3. Note the tendency for the isotherms (lines of constant temperature) to follow the latitude belts, especially in the Southern Hemisphere, and for the temperature to decrease in value toward the poles. This is a result of the decrease in solar radiation with latitude. The influence of ocean currents can be seen in the distortion of the isotherms. In the western North Atlantic Ocean there is a definite north-ward bowing of the 15, 10, 5, and 0°C sea temperature contours in the region of the warm, northward-flowing Gulf Stream. In the western South Atlantic the warm Brazil Current, flowing along the east coast of South America, influences the location of the 25°C isotherm, while in the eastern South Atlantic the effect of the cold Benguela Current flowing north along the west coast of Africa can be seen. The sea–air temperature differences at various locations in the Northern (winter) Hemisphere are greater than in the Southern Hemisphere. This indicates the strong heat exchange between sea and air that occurs in the winter compared with the summer.

The variables of state in both the oceans and the atmosphere exhibit a

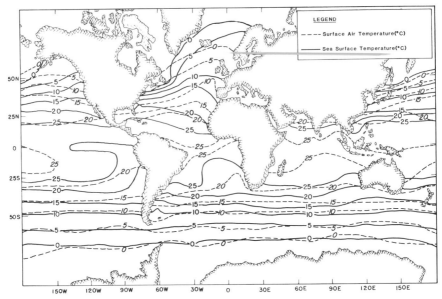

Figure 1.3. Average surface air and sea temperature (°C) for January. (Adapted from U.S. Navy *Marine Climatic Atlas of the World*, Volumes I–VII, 1965 to 1978, and NAVPERS, 1965.)

strong tendency toward vertical stratification, or layering. The atmosphere, heated mainly from below, cools at an average "lapse" rate of about 6.5°C/ km to the top of the troposphere. The troposphere is the lowest thermal layer in the atmosphere and extends upward about 17 km near the equator, 11 km in mid-latitudes, and 9 km in polar regions. The tropopause, a region of relatively constant temperature, separates the upper troposphere from the stratosphere. Nearly all weather processes occur below the tropopause.

The oceans, heated mainly at their upper boundary, cool with increasing depth. Surface winds generally mix the upper 100 to 200 m, forming the isothermal surface mixed layer. Below this layer temperature decreases rapidly in the main thermocline, a region typically 500 to 1000 m deep in mid-latitudes. In the oceanic deep layer below the thermocline temperature decreases slowly, reaching values as low as 2 to $-1°C$ at depths below 3000 m. Cold, dense water of polar origin is found at the bottom of all ocean basins. Figure 1.4 illustrates the average vertical distribution of temperature for both sea and air at a mid-latitude location in the winter.

The second variable of state, pressure, is a measure of the force exerted on a unit area at a given level within the fluid. A previous discussion outlined how the surface atmospheric pressure changes horizontally and how this is related to the global circulation. Pressure, like temperature, also varies over the vertical direction in both the sea and air. The pressure at any level can be

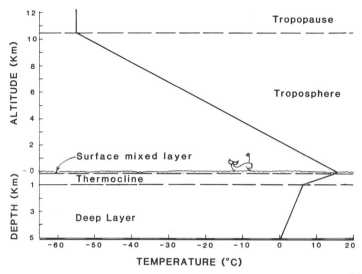

Figure 1.4. Average thermal structure of the atmosphere and the ocean at a mid-latitude location in winter.

related to the weight of the fluid above it by the hydrostatic equation

$$p = \bar{\rho}gH$$

where p is the hydrostatic air or water pressure, $\bar{\rho}$ is the fluid density averaged over the depth H, and g is gravitational acceleration (980 cm/s²). When a typical value for average density within the bottom 20 km of the atmosphere is substituted in the hydrostatic equation, the surface pressure can be calculated as

$$p = (0.5 \times 10^{-3} \text{ g/cm}^3)(980 \text{ cm/s}^2)(20 \times 10^5 \text{ cm})$$

or

$$p \simeq 10^6 \text{ dyne/cm}^2$$

This value, $p = 10^6$ dyne/cm², called a bar, represents the approximate surface atmospheric pressure. Since changes in pressure are usually small compared to the bar, the millibar (10^3 dyne/cm²) is used by meteorologists to report air-pressure values. Oceanic pressure is expressed in decibars (10^5 dyne/cm²) because of the relationship between oceanic pressure in decibars and ocean depth expressed in meters. Using the hydrostatic equation and a mean sea water density $\bar{\rho} = 1.02$ g/cm^3, we can calculate the pressure at a depth of 1 m as

$$p = (1.02 \text{ g/cm}^3)(980 \text{ cm/s}^2)(10^2 \text{cm})$$

or

$$p \simeq 10^5 \text{ dyne/cm}^2$$

Thus the depth in meters is numerically very close in value to the pressure in decibars. The MKS unit of pressure (the pascal) is also used in meteorology and oceanography.

On both atmospheric and oceanic pressure charts lines of equal pressure are called isobars.

The third physical parameter that is important to an understanding of the state of the atmosphere is water substance. This is defined as atmospheric water in solid form (snow or hail), liquid form (clouds or rain), or vapor form (invisible water vapor). The global-scale wind and pressure distribution at the Earth's surface is directly related to the global distribution of water substance. For example, the ITCZ represents a surface low-pressure region within which large amounts of warmed, moist air rise into the upper atmosphere. The subsequent condensation at cooler altitudes produces a global-scale cloud and rain pattern observed in equatorial latitudes. Similarly, along the polar fronts, which are located in low-pressure belts at about 60°N and 60°S latitudes, rising air forms widespread cloudiness and precipitation. By contrast, in the global high-pressure regions air is slowly sinking toward the Earth's surface. This subsidence tends to produce regions with clear skies and minimum precipitation.

The third variable of state in the ocean is salinity, which is defined as the amount of dissolved material in a kilogram of sea water when all of the carbonate has been converted to oxide, all of the bromine and iodine replaced by chlorine, and all of the organic matter oxidized. In simple terms, salinity is the mass (in grams) of dissolved material in one kilogram of sea water. Salinity is related to chlorinity, a more easily measured constituent of sea water, by the empirical formula

$$S(\text{‰}) = 1.8066 \text{ Cl}(\text{‰})$$

where $S(\text{‰})$ is salinity expressed in parts per thousand and $Cl(\text{‰})$ is chlorinity, also expressed in parts per thousand. With present-day oceanographic instrumentation salinity is determined from measurements of the electrical conductivity of sea water.

On a global scale the ocean surface salinity pattern is related to the surface atmospheric pressure field and to the distribution of atmospheric water substance. In regions of high precipitation, such as low-pressure areas of the

ITCZ and the polar fronts, the ocean surface salinity is relatively low (approximately 34 to 35 ‰). In ocean areas near the subtropical high-pressure zones, ocean surface salinity is generally at a maximum (approximately 36 to 37 ‰), due to the extensive evaporation and low amounts of precipitation.

Information on the distribution of the variables of state is also used in conjunction with the appropriate equation of state to calculate the density of the sea water or air. Density is defined as the mass of a substance per unit volume. The standard dry-air surface density for a pressure of 1013.25 millibars and temperature of 0°C is $\rho_a = 1.29 \times 10^{-3}$ g/cm^3. Air density is relatively insensitive to the typical atmospheric variations in water substance.

In the ocean, density is calculated from observations of temperature, pressure, and salinity. Sea water is slightly denser than fresh water (about 1.025 g/cm^3, compared with 1.000 g/cm^3), but some 800 times denser than air. The range of sea water density is from about 1.020 to 1.030 g/cm^3. The greatest changes in sea water density occur near the sea surface and near coastlines. Density can be decreased by precipitation, river runoff, ice melt, and solar heating. As sea water becomes less dense, it tends to float over denser water. This is a dynamically stable condition in which there is little tendency for the water to mix. Sea water density can also be increased, by evaporation, formation of sea ice, and cooling. If the surface water becomes denser than the underlying water, it will sink to the level of equal density. Less dense water will then rise to make room for the sinking water. This leads to a convective circulation that continues until the density field becomes stable again. In polar regions cold surface waters continuously sink toward the ocean floor, eventually spreading across the bottom of all the world's ocean basins. Variations in salinity also cause changes in sea water density. For example, the low-salinity ocean waters near the ITCZ are less dense than the higher-salinity waters near the subtropical high-pressure zone.

Sea water density is generally calculated to five decimal places. For convenience a term called sigma, $\sigma_{s,t,p}$, has been defined as

$$\sigma_{s,t,p} = (\rho_{s,t,p} - 1) \times 10^3$$

where $\rho_{s,t,p}$ is the water density calculated at a specific salinity, temperature, and pressure. For example, a $\rho_{s,t,p}$ value of 1.0255 g/cm^3 is equal to a sigma value of 25.5 [$\sigma_{s,t,p} = (1.0255 - 1) \times 10^3 = 25.5$]. In the upper layers of the ocean the effect of pressure on density is small. Accordingly, a second parameter called sigma-t, σ_t, has been defined that accounts for variations of density due to temperature and salinity changes only. Sigma-t is defined as

$$\sigma_t = \sigma_{s,t,0} = (\rho_{s,t,0} - 1) \times 10^3$$

where the pressure is taken as surface atmospheric pressure (considered zero).

1.5. AIR MASSES AND WATER MASSES

Both synoptic and climatic charts often show large regions of relatively constant density separated by rapid-transition zones. In the atmosphere, these vast regions of relatively constant temperature and water-vapor content are called air masses. In the ocean, similar vast regions of relatively constant temperature and salinity are called water masses. The transition zones that separate these constant density regions are called fronts. Figure 1.5 shows the global distribution of the major air-mass source regions in January. A source region is defined as a large geographic area over which the air mass eventually takes on the local temperature and water-vapor characteristics. Air masses formed over the polar high-pressure regions are therefore cold and dry, whereas those formed near the equator are warm. Air masses formed over the oceans have higher moisture content than air masses formed over land. Air masses can be identified by plotting the atmospheric variables of state. Figure 1.6 is an example of a temperature versus water-vapor chart plotted at various atmospheric pressures. The chart identifies three distinct air masses. A cold, dry continental polar (cP) air mass; a cold, moist maritime polar (mP) air mass; and a warm, moist maritime tropical (mT) air mass are shown as three separate lines on the chart. In addition to showing air masses, daily synoptic weather charts often show smaller-scale elliptical high- and low-pressure systems ranging in size from hundreds to thousands of kilometers in diameter. The high-pressure systems, also called anticyclones, are fair-weather regions with winds circulating clockwise about their centers in the Northern Hemisphere and counterclockwise in the Southern Hemisphere. Low-pressure systems, also called cyclones, are often regions of clouds and precipitation. They have counterclockwise winds about their centers in the Northern Hemisphere and clockwise winds in the Southern Hemisphere.

Water masses, like their meteorological counterparts, take on the temperature and salinity characteristics of their source region. Thus, water masses formed in equatorial waters are warm with relatively low salinities, whereas those formed in the subtropical oceans are warm with relatively higher salinities. Water masses can also be identified by the use of plots of the variables of state. An example of this is the T/S diagram illustrated in Figure 1.7. On this chart polar, equatorial, and Mediterranean water masses can be identified on the basis of temperature and salinity values at various subsurface depths (pressures). Each water mass can be identified by a line that converges to a common point. This particular T/S diagram also illustrates the common origin of all ocean bottom waters in the cold Arctic or Antarctic

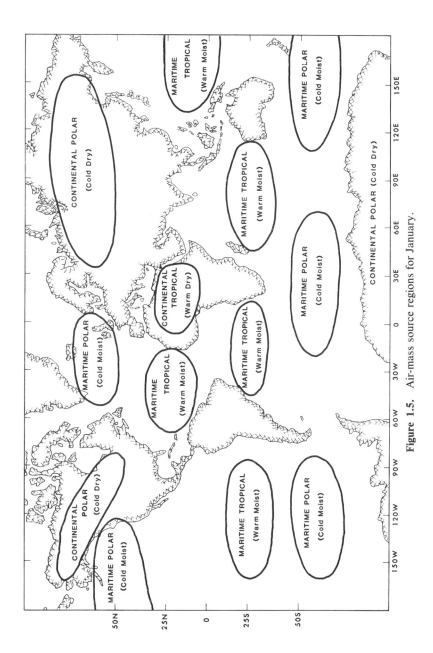

Figure 1.5. Air-mass source regions for January.

13

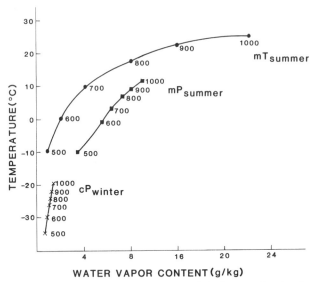

Figure 1.6. Temperature (°C) versus water vapor (g/kg) diagram showing three air masses: continental Polar (cP) in winter (x), maritime Tropical (mT) in summer (●), and maritime Polar (mP) in summer (■). The numbers on the chart indicate atmospheric pressure in millibars.

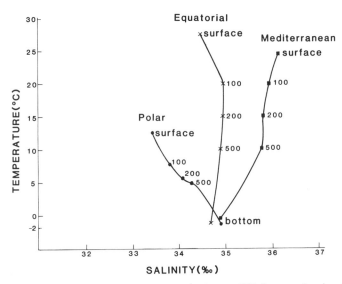

Figure 1.7. Temperature (°C) versus salinity (‰) chart or T/S diagram showing three water masses: Polar (●), Equatorial (x), and Mediterranean (■). The numbers on the chart indicate ocean pressure in decibars.

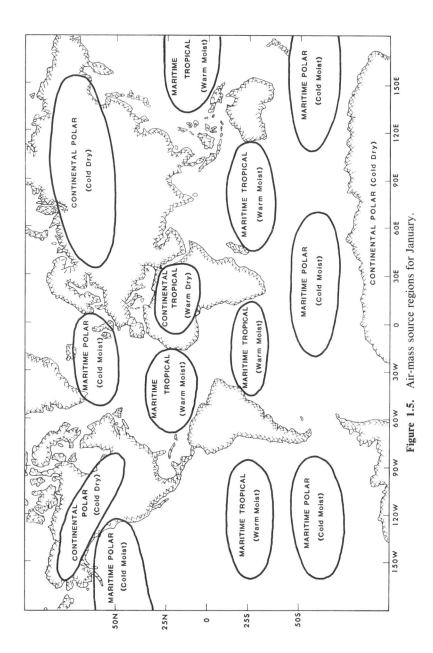

Figure 1.5. Air-mass source regions for January.

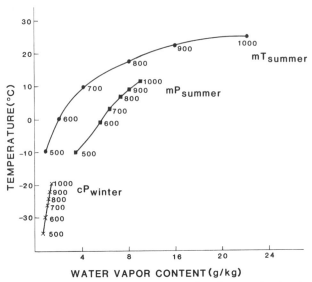

Figure 1.6. Temperature (°C) versus water vapor (g/kg) diagram showing three air masses: continental Polar (cP) in winter (x), maritime Tropical (mT) in summer (●), and maritime Polar (mP) in summer (■). The numbers on the chart indicate atmospheric pressure in millibars.

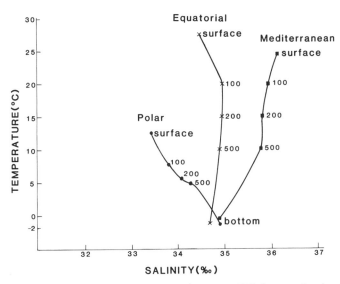

Figure 1.7. Temperature (°C) versus salinity (‰) chart or T/S diagram showing three water masses: Polar (●), Equatorial (x), and Mediterranean (■). The numbers on the chart indicate ocean pressure in decibars.

waters, since the common point where the lines converge represents cold polar water.

1.6. ATMOSPHERIC AND OCEANIC FRONTS

Fronts are dynamically active zones where rapid changes in the variables of state and velocity are observed. Meteorological fronts separate air masses. On a global scale the polar fronts separate the polar and subtropical high-pressure regions. As one crosses a polar front from the subtropical high-pressure region toward the polar high, temperature decreases rapidly and the wind changes direction. In addition to changing in the horizontal dimension, wind speed also increases with altitude above the surface location of the polar front throughout the troposphere up to the tropopause. The result of this wind shear, or variation of wind with altitude, is a strong upper atmospheric wind known as the jet stream. This wind, located near the tropopause, sometimes exceeds speeds of 100 m/s.

The boundary between two water masses is an oceanographic front. In this region rapid changes in water temperature, salinity, and current velocity are observed. As with its meteorological counterpart, the strong horizontal density changes that accompany oceanic fronts are related dynamically to current shear and strong, relatively narrow currents commonly observed near the frontal zone. Figure 1.8 shows a vertical temperature cross section for the Gulf Stream front. The cold coastal and warm Gulf Stream water masses are easily identified. The front itself, known as the North Wall, is shown separating two water masses. The weaker Slope front is also shown separating the nearshore coastal water from the offshore Slope water. The distinct downward sloping tendency of the isotherms near the North Wall is characteristic of a frontal zone, and is dynamically related to the Gulf

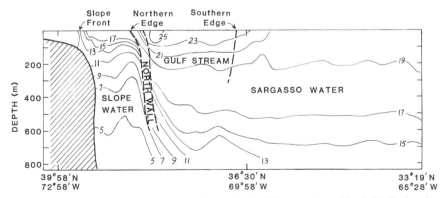

Figure 1.8. Temperature (°C) cross section of the Gulf Stream front (North Wall) showing Slope, Gulf Stream, and Sargasso water masses. (Adapted from U.S. Navy *Tactical Applications Guide*, 1981.)

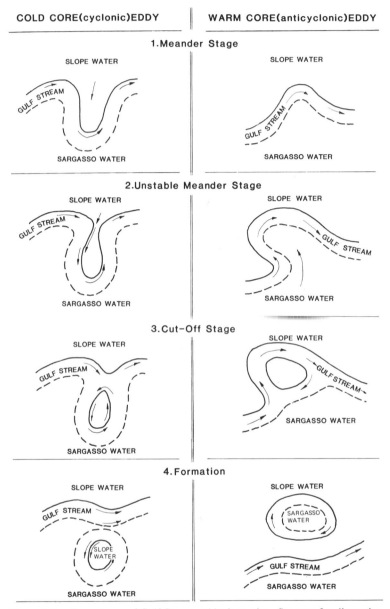

Figure 1.9.   Four stages of Gulf Stream eddy formation. See text for discussion.

Stream, a rapid current about 100 to 150 km across that is the oceanic counterpart to the jet stream. Current speeds of up to 100 to 200 cm/s are found in the high-velocity core of the Gulf Stream.

1.7. OCEAN EDDIES

Associated with the Gulf Stream, and with many other major ocean current systems, are nearly circular anticyclonic and cyclonic eddies with diameters of 100 to 200 km and current speeds of up to 100 and 200 cm/s. These eddies often result from large unstable meanders of the frontal boundary. Gulf Stream anticyclonic eddies have warm water at their center, while cyclonic Gulf Stream eddies have a cold core. Once formed, these closed circulation systems have a life span of 3 to 6 months or even longer. Figure 1.9 illustrates the four stages of formation of both the warm- and cold-core eddies that develop along the Gulf Stream front. Warm-core eddies form when extreme northward frontal meanders produce a closed clockwise gyre with a center of warm Sargasso water surrounded by a ring of Gulf Stream water.

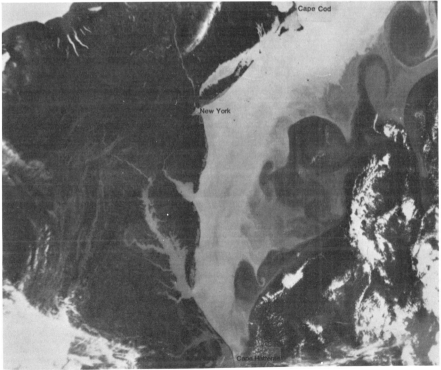

Figure 1.10. Thermal pattern of the Gulf Stream current system as shown on a satellite image from May 1, 1977. (Courtesy of National Oceanic and Atmospheric Administration.)

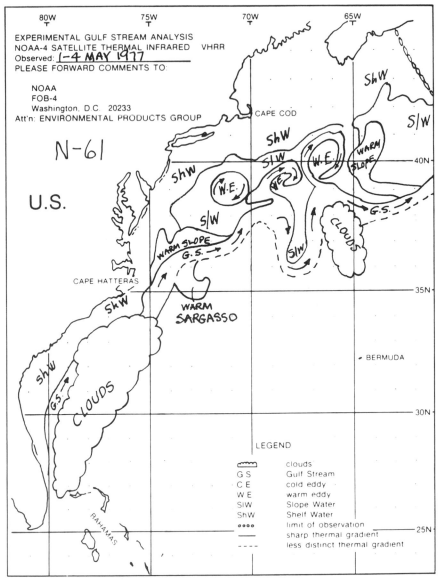

Figure 1.11. Gulf Stream frontal analysis chart based on the satellite image shown in Figure 1.10. (Courtesy of National Oceanic and Atmospheric Administration.)

Cold-core eddies result from extreme southward meanders of the frontal boundary forming a closed counterclockwise gyre around the cold Slope water.

Gulf Stream eddies form east of 70°W longitude and then move slowly southwestward, usually remaining near the frontal boundary. Warm-core eddies generally rejoin the Gulf Stream in the shallow waters off Cape Hatteras, North Carolina, whereas cold-core eddies move along the outer edge of the Gulf Stream and have been observed far south of Cape Hatteras. Similar eddy structures are produced near other boundary currents throughout the world's oceans.

1.8. OPERATIONAL PRODUCTS BASED ON SATELLITE IMAGERY

Charts of oceanic fronts and eddies can be produced in near real time using available satellite images. The strong thermal gradients at the sea surface that are typical of these oceanic features lead to distinct gray patterns on satellite images that allow detailed charts of fronts and eddies to be produced. Oceanographers can combine local climatological knowledge about frontal currents with these images to make real-time charts of surface currents.

One example of such a product is the Gulf Stream frontal analysis chart produced by the National Oceanic and Atmospheric Administration (NOAA). The analysis uses polar orbiting satellite imagery with a resolution of 1 km. Figure 1.10 is an example of a surface thermal pattern observed by satellite for the Gulf Stream region, showing coastal water, the Gulf Stream front, and several warm-core eddies. The warm Gulf Stream (dark region) is the meandering ribbon of water between the cold coastal water (light gray) and the very warm Sargasso water.

In the NOAA analysis procedure a plastic latitute–longitude grid is positioned over the satellite image so that ocean fronts and eddies can be mapped and eventually transferred to a base chart. Generally, a number of images are combined to give a detailed picture of the frontal zone and eddies in the map area. Sea surface temperatures from merchant ships and data from fixed buoys are added to the analysis. An example of the final product is shown in Figure 1.11. The solid lines represent the observed frontal features and dashed lines represent estimated features. Water masses and eddies are also charted.

This Gulf Stream analysis chart has been used for numerous applications, including planning for search and rescue operations, locating likely areas of fish concentrations, calculating trajectories for oil spills, and optimum routing of deep-draft tankers.

A SURVEY OF OCEAN
CURRENT DYNAMICS

Ocean currents influence physical, chemical, and biological processes that occur over the complete spectrum of oceanic time and space. In this chapter a framework is developed that relates ocean currents to the governing equations of fluid flow. Topics addressed include the equations of motion, the continuity equation, and several simplifications of these equations that lead to useful dynamical relationships. The intent is to survey basic ocean dynamics using examples that emphasize the application of theory to specific problems.

2.1. THE EQUATION OF MOTION

Fluid flow in the ocean is governed by Newton's second law of motion and by the principle of the conservation of mass. Newton's law,

$$\Sigma \mathbf{F} = m\mathbf{a}$$

relates the vector sum of the forces, $\Sigma \mathbf{F}$, acting on a fluid parcel to its mass m and its acceleration vector $\mathbf{a}$. A vector quantity has both magnitude and direction. The principal forces acting on a fluid parcel are the pressure gradient force, the Coriolis force, the gravitational force, the frictional force, and the centrifugal force. These forces are expressed in Newton's law in terms of accelerations given as

$$\mathbf{a} = \frac{\Sigma \mathbf{F}}{m}$$

$$\frac{d\mathbf{V}}{dt} = \frac{\mathbf{PGF}}{m} + \frac{\mathbf{CF}}{m} + \frac{\mathbf{G}}{m} + \frac{\mathbf{F}}{m} + \frac{\mathbf{C}}{m}$$

where $\mathbf{V}$ = the velocity vector, which represents the speed and direction with which the fluid parcel is moving

$\frac{d\mathbf{V}}{dt}$ = the acceleration vector, which represents the parcel's change in velocity with time

$\frac{\mathbf{PGF}}{m}$ = the pressure gradient force per unit mass

$\frac{\mathbf{CF}}{m}$ = the Coriolis force per unit mass

$\frac{\mathbf{G}}{m}$ = the gravitational force per unit mass

$\frac{\mathbf{F}}{m}$ = the frictional force per unit mass

$\frac{\mathbf{C}}{m}$ = the centrifugal force per unit mass.

This relationship, called the equation of motion, is one of the two basic equations of ocean dynamics. The vector equation of motion may be expressed in component form with respect to a given coordinate system. In a left-handed Cartesian coordinate system, which is used in oceanography, the x-axis is positive to the east, the y-axis is positive to the north, and the z-axis is positive downward into the ocean. In a right-handed Cartesian coordinate system, also used in meteorology and oceanography, the z-axis is positive upward. In these coordinate systems, u represents the velocity component in the x-direction, v the velocity component in the y-direction, and w the velocity component in the z-direction. In the remainder of this section each term in the component equations of motion will be examined as it appears in a Cartesian coordinate system.

2.1.1.   The Acceleration Term

Forces acting on a fluid parcel produce changes in its direction and speed. Such changes, or accelerations, occurring at a specific place in a fluid over a given time period are called local accelerations, while changes occurring as the fluid parcel is being transported by the fluid flow are called advective accelerations. The total acceleration is the sum of local and advective terms. In component form total acceleration has an x-component:

$$\underset{\text{Total}}{\underbrace{}} = \underset{\text{Local}}{\underbrace{}} + \overset{\text{Advective}}{\overbrace{}}$$

$$\frac{du}{dt} = \frac{\partial u}{\partial t} + u\,\frac{\partial u}{\partial x} + v\,\frac{\partial u}{\partial y} + w\,\frac{\partial u}{\partial z}$$

a *y*-component:

$$\underset{\text{Total}}{} = \underset{\text{Local}}{} + \overset{\text{Advective}}{\overbrace{}}$$

$$\frac{dv}{dt} = \frac{\partial v}{\partial t} + u\,\frac{\partial v}{\partial x} + v\,\frac{\partial v}{\partial y} + w\,\frac{\partial v}{\partial z}$$

and a *z*-component:

$$\underset{\text{Total}}{} = \underset{\text{Local}}{} + \overset{\text{Advective}}{\overbrace{}}$$

$$\frac{dw}{dt} = \frac{\partial w}{\partial t} + u\,\frac{\partial w}{\partial x} + v\,\frac{\partial w}{\partial y} + w\,\frac{\partial w}{\partial z}$$

In some applications of the equations of motion the fluid is considered to be nonaccelerated so that

$$\frac{du}{dt} = \frac{dv}{dt} = \frac{dw}{dt} = 0$$

In other applications of the equations of motion the fluid flow is considered to be steady. This means that the local acceleration vanishes, or

$$\frac{\partial u}{\partial t} = \frac{\partial v}{\partial t} = \frac{\partial w}{\partial t} = 0$$

2.1.2. Pressure Gradient Force per Unit Mass

Horizontal changes in pressure occur in conjunction with high- and low-pressure regions in both the atmosphere and oceans. Fluid parcels flow from the high- toward the low-pressure areas. The greater the pressure change over a given distance the faster the flow. This change in pressure over distance, called the pressure gradient, is an important driving mechanism for both winds and currents. The pressure gradient force (*PGF*) per unit mass, which is an acceleration, is given in component form as an *x*-component:

$$\left(\frac{PGF}{m}\right)_x = -\frac{1}{\rho}\frac{\partial p}{\partial x}$$

a y-component:

$$\left(\frac{PGF}{m}\right)_y = -\frac{1}{\rho}\frac{\partial p}{\partial y}$$

and a z-component

$$\left(\frac{PGF}{m}\right)_z = -\frac{1}{\rho}\frac{\partial p}{\partial z}$$

where ρ is the fluid density and the minus sign indicates flow toward regions of lower pressure.

2.1.3. Coriolis Force per Unit Mass

The effect of the Earth's rotation is included in the equations of motion through the Coriolis force. This term is necessary because the coordinate system used to described geophysical fluid flow at a fixed place on Earth is continuously moving as the Earth rotates. The Coriolis force (CF) per unit mass generally appears in component form as an x-component:

$$\left(\frac{CF}{m}\right)_x = fv$$

a y-component:

$$\left(\frac{CF}{m}\right)_y = -fu$$

and a z-component:

$$\left(\frac{CF}{m}\right)_z = 0$$

where $f = 2\,\Omega\,\sin\phi$ is called the Coriolis parameter; $\Omega = 7.29 \times 10^{-5}$ rad/s is the angular velocity of the Earth's rotation, and ϕ is the parcel's geographic latitude.

The Coriolis force causes fluid parcels to be deflected to the right of the flow direction in the Northern Hemisphere and to the left in the Southern Hemisphere, with a magnitude dependent on the speed of the flow and the location (latitude) of the fluid parcel. The force vanishes if the fluid parcel is at rest or at the equator, and is a maximum at the poles. The Coriolis deflection plays a major role in the modification of the large-scale circulation of both the atmosphere and ocean.

2.1.4. Gravitational Force per Unit Mass

Fluid parcels are pulled toward the Earth's center of mass by gravitational attraction. On the rotating Earth there is also a small component of centrifugal force directed outward from the axis of rotation. The sum of these two forces is called gravity (G). The magnitude of the gravitational force changes over the Earth because of slight variations in the distance from the Earth's surface to its center of mass and because of changes in mass from place to place. In most geophysical problems the acceleration due to gravity is expressed in component form as an x-component:

$$\left(\frac{G}{m}\right)_x = 0$$

a y-component:

$$\left(\frac{G}{m}\right)_y = 0$$

and a z-component:

$$\left(\frac{G}{m}\right)_z = g = 980 \text{ cm/s}^2$$

2.1.5. Frictional Force per Unit Mass

Friction slows fluid flow and generally appears in the equations of motion as a stress term. In the atmosphere and the ocean turbulent eddies embedded in the mean flow are the mechanism by which frictional effects are spread throughout the fluid. In the following approach, known as the Reynolds' formulation for the equations of motion, the instantaneous velocity components are written as

$$u = \bar{u} + u'$$
$$v = \bar{v} + v'$$
$$w = \bar{w} + w'$$

where $\bar{u}$, $\bar{v}$, and $\bar{w}$, are the velocity components averaged over a given time period at a particular point in space. The time period must be long enough that the effects of random turbulent eddies in the flow field are smoothed out. The quantities u', v', and w' are instantaneous fluctuations from the mean values caused by these turbulent eddies.

The equations of motion for the mean flow can be derived by averaging the instantaneous equations over time. When this is done, a number of additional terms resulting from the averaging process enter the equations. These terms represent the turbulent mixing effects of the eddies and take the form of frictional forces (F) per unit mass. In component form these frictional accelerations are given as an x-component:

$$\left(\frac{F}{m}\right)_x = -\frac{1}{\rho}\left[\frac{\partial}{\partial x}\,\rho(\overline{u'u'}) + \frac{\partial}{\partial y}\,\rho(\overline{u'v'}) + \frac{\partial}{\partial z}\,\rho(\overline{u'w'})\right]$$

a y-component:

$$\left(\frac{F}{m}\right)_y = -\frac{1}{\rho}\left[\frac{\partial}{\partial x}\,\rho(\overline{v'u'}) + \frac{\partial}{\partial y}\,\rho(\overline{v'v'}) + \frac{\partial}{\partial z}\,\rho(\overline{v'w'})\right]$$

and a z-component:

$$\left(\frac{F}{m}\right)_z = -\frac{1}{\rho}\left[\frac{\partial}{\partial x}\,\rho(\overline{w'u'}) + \frac{\partial}{\partial y}\,\rho(\overline{w'v'}) + \frac{\partial}{\partial z}\,\rho(\overline{w'w'})\right]$$

where the symbol $\overline{u'u'}$ indicates the averaged value of $u'u'$, and $\bar{u}' = \bar{v}' = \bar{w}' = 0$.

The terms in the above equations can be arranged as an array of nine stresses collectively called a stress tensor. The elements of the stress tensor are:

$$\tau_{xx} = -\rho(\overline{u'u'})\;;\; \tau_{xy} = -\rho(\overline{u'v'})\;;\; \tau_{xz} = -\rho(\overline{u'w'})$$

$$\tau_{yx} = -\rho(\overline{v'u'})\;;\; \tau_{yy} = -\rho(\overline{v'v'})\;;\; \tau_{yz} = -\rho(\overline{v'w'})$$

$$\tau_{zx} = -\rho(\overline{w'u'})\;;\; \tau_{zy} = -\rho(\overline{w'v'})\;;\; \tau_{zz} = -\rho(\overline{w'w'})$$

These elements, called the Reynolds' stresses, are difficult to measure, and so are generally inferred from observations of the mean velocity. Prandtl (1952) related the Reynolds' stress to the gradient in the mean velocity components using a turbulent-eddy viscosity coefficient such that

$$\tau_{xx} = K_x\frac{\partial \bar{u}}{\partial x}\;;\; \tau_{xy} = K_y\frac{\partial \bar{u}}{\partial y}\;;\; \tau_{xz} = K_z\frac{\partial \bar{u}}{\partial z}$$

$$\tau_{yx} = K_x\frac{\partial \bar{v}}{\partial x}\;;\; \tau_{yy} = K_y\frac{\partial \bar{v}}{\partial y}\;;\; \tau_{yz} = K_z\frac{\partial \bar{v}}{\partial z}$$

$$\tau_{zx} = K_x\frac{\partial \bar{w}}{\partial x}\;;\; \tau_{zy} = K_y\frac{\partial \bar{w}}{\partial y}\;;\; \tau_{zz} = K_z\frac{\partial \bar{w}}{\partial z}$$

where K_x, K_y, and K_z, are the eddy viscosity coefficients with dimensions M/LT. In the ocean turbulent friction is much greater than friction between molecules, so molecular friction is generally neglected in the equations of motion for the mean flow. The commonly used assumption of a constant eddy viscosity coefficient leads to frictional acceleration terms in the equation of motion of the form

$$\frac{1}{\rho}\left[\frac{\partial}{\partial x} - \rho(\overline{u'u'})\right] = \frac{1}{\rho}\frac{\partial \tau_{xx}}{\partial x} = \frac{K_x}{\rho}\frac{\partial^2 \bar{u}}{\partial x^2}$$

2.1.6. Centrifugal Force per Unit Mass

In addition to causing the Coriolis force, the Earth's rotation results in a centrifugal force which acts outward from the Earth's axis of rotation. The vertical component of this force is incorporated into the gravity term, while the horizontal component is generally neglected because of its relatively small magnitude compared with other forces in the equations of motion. In smaller-scale circular motions, for example, the flow around oceanic eddies, the centrifugal force (C) also enters the equations of motion in the form

$$\frac{C}{m} = \frac{V^2}{R}$$

where V is the fluid parcel's speed and R is the radius of its circular path.

2.1.7. The Component Equations of Motion

The complete averaged equations of motion can now be written as an x-component (east):

$$\frac{\partial \bar{u}}{\partial t} + \bar{u}\frac{\partial \bar{u}}{\partial x} + \bar{v}\frac{\partial \bar{u}}{\partial y} + \bar{w}\frac{\partial \bar{u}}{\partial z} =$$

$$-\frac{1}{\rho}\frac{\partial \bar{p}}{\partial x} + f\bar{v} + \frac{K_x}{\rho}\frac{\partial^2 \bar{u}}{\partial x^2} + \frac{K_y}{\rho}\frac{\partial^2 \bar{u}}{\partial y^2} + \frac{K_z}{\rho}\frac{\partial^2 \bar{u}}{\partial z^2}$$

a y-component (north):

$$\frac{\partial \bar{v}}{\partial t} + \bar{u}\frac{\partial \bar{v}}{\partial x} + \bar{v}\frac{\partial \bar{v}}{\partial y} + \bar{w}\frac{\partial \bar{v}}{\partial z} =$$

$$-\frac{1}{\rho}\frac{\partial \bar{p}}{\partial y} - f\bar{u} + \frac{K_x}{\rho}\frac{\partial^2 \bar{v}}{\partial x^2} + \frac{K_y}{\rho}\frac{\partial^2 \bar{v}}{\partial y^2} + \frac{K_z}{\rho}\frac{\partial^2 \bar{v}}{\partial z^2}$$

and a z-component (downward):

$$\frac{\partial \bar{w}}{\partial t} + \bar{u}\,\frac{\partial \bar{w}}{\partial x} + \bar{v}\,\frac{\partial \bar{w}}{\partial y} + \bar{w}\,\frac{\partial \bar{w}}{\partial z} =$$

$$-\frac{1}{\rho}\,\frac{\partial \bar{p}}{\partial z} + g + \frac{K_x}{\rho}\,\frac{\partial^2 \bar{w}}{\partial x^2} + \frac{K_y}{\rho}\,\frac{\partial^2 \bar{w}}{\partial y^2} + \frac{K_z}{\rho}\,\frac{\partial^2 \bar{w}}{\partial z^2}$$

These equations are generally written without the bars over the various quantities, although the averaging process is still implied. The averaged equations of motion form the physical basis for a number of calculations related to ocean currents.

2.2. THE EQUATION OF CONTINUITY

The second important relationship governing fluid flow in the ocean is derived from the law of conservation of mass, and states that the total rate of change of density in an infinitesimal fluid volume of fixed size is proportional to the spatial change in velocity. The equation of continuity expresses this relationship as

$$-\frac{1}{\rho}\,\frac{d\rho}{dt} = \frac{\partial u}{\partial x} + \frac{\partial v}{\partial y} + \frac{\partial w}{\partial z}$$

where the term on the right-hand side of this equation is the three-dimensional velocity divergence. After rearranging terms, the continuity equation can also be written as

$$-\frac{\partial \rho}{\partial t} = \left[\frac{\partial}{\partial x}\,(\rho u) + \frac{\partial}{\partial y}\,(\rho v) + \frac{\partial}{\partial z}\,(\rho w) \right]$$

which relates the local rate of change of density to the three-dimensional mass divergence. If density is assumed to be constant the three-dimensional velocity divergence is zero, or

$$\frac{\partial u}{\partial x} + \frac{\partial v}{\partial y} + \frac{\partial w}{\partial z} = 0$$

The three component equations of motion are combined with the continuity equation to give four mathematical relationships describing ocean fluid flow. In addition, an empirical equation of state relates the variables of state (temperature, salinity, and pressure) to density. These five relationships form a closed set of equations for the five unknown variables u, v, w, p, and ρ. In theory, given initial and boundary conditions, the equations can be solved to describe the fluid flow completely. In practice, simplified forms of these equations that approximate real oceanic conditions have provided useful results.

2.3.   FRICTIONLESS NONACCELERATED FLOW

The equations of motion provide the theoretical basis for the calculation of winds and currents. One simplified form of these equations assumes nonaccelerated frictionless flow. The component equations of motion can then be written as an x-component:

$$fv = \frac{1}{\rho} \frac{\partial p}{\partial x}$$

a y-component:

$$fu = -\frac{1}{\rho} \frac{\partial p}{\partial y}$$

and a z-component:

$$g = \frac{1}{\rho} \frac{\partial p}{\partial z}$$

The first two equations represent a balance between the Coriolis and pressure gradient forces called geostrophic flow. The geostrophic equations have classically been used to estimate winds or currents based on measurements of the horizontal pressure distribution.

To better understand geostrophic flow, consider the following. The Coriolis force is zero for a fluid parcel at rest. As the parcel is accelerated toward regions of lower pressure in a horizontal pressure field, it will be deflected more and more to the right in the Northern Hemisphere and to the left in the Southern Hemisphere. Eventually it will turn perpendicular to its original trajectory so that flow is parallel to the isobars (lines of constant pressure). Figure 2.1 illustrates the process that produces geostrophic equilibrium for both hemispheres. The "Law of Storms" used for years by mariners is based on the principle of geostrophic balance. This law states that in the Northern Hemisphere the geostrophic wind, which is a good approximation to the actual wind under many circumstances, blows parallel to the isobars with low pressure to the left of an observer whose back is to the flow. The flow direction is opposite to this in the Southern Hemisphere

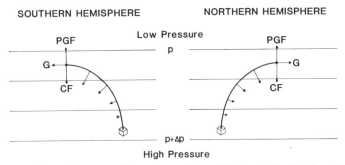

Figure 2.1. Fluid parcels approaching geostrophic equilibrium in the Northern Hemisphere and in the Southern Hemisphere. In both cases **G** represents the geostrophic flow, **PGF** is the pressure gradient force, and **CF** is the Coriolis force.

due to the opposite direction in which the Coriolis force acts in the Southern Hemisphere.

The z-component equation represents a balance between gravity and the vertical pressure gradient, which are the primary vertical forces. This balance is called hydrostatic equilibrium. The hydrostatic equation is often used to calculate pressure at a given depth in the ocean or at a given altitude in the atmosphere. The geostrophic and hydrostatic simplifications of the equations of motion have been especially useful in the solution of large-scale meteorological and oceanographic problems in which fluid acceleration and frictional effects can be assumed to be relatively small. For example, large-scale ocean currents are approximately in geostrophic equilibrium. Therefore, information on the pressure gradient allows calculations to be made of these currents. The oceanic pressure gradient, represented by the slope of the isobaric surfaces, is usually inferred from shipboard measurements of temperature and salinity. This slope can also be measured by remote sensing techniques. The slope of the sea surface, which is the uppermost isobaric level in the ocean, can be used for geostrophic current calculations of, for example, the Gulf Stream. This current is generally about 100 km wide, across which distance the sea surface has been observed to rise about 1 m in the offshore direction. Using the geostrophic equation, with the x-axis perpendicular to the current pointing offshore, the current speed V_{GS} is given as

$$V_{GS} = \frac{1}{\rho f} \frac{\Delta p}{\Delta x}$$

where Δp is the change in pressure across the Gulf Stream due to the increase in sea level and $\Delta x = 100$ km is the current width.

From the hydrostatic equation,

$$\Delta p = \rho g \Delta z$$

where $\Delta z = 1$ m is the rise in sea level across the current. The estimated geostrophic current magnitude is therefore

$$V_{GS} = \frac{g}{f} \frac{\Delta z}{\Delta x} = \left(\frac{980 \text{ cm/s}^2}{10^{-4} \text{ rad/s}} \right) \left(\frac{10^2 \text{ cm}}{10^7 \text{ cm}} \right)$$

$$V_{GS} \simeq 100 \text{ cm/s}$$

where the slope of the sea surface $= 10^{-5}$, $g = 980$ cm/s^2, and $f = 10^{-4}$ rad/s. The current flows in the positive y-direction (approximately northward) parallel to the isobaric surfaces, with the lowest sea surface elevation to the left (Northern Hemisphere) if the observer's back is to the current. In the future, oceanographers will use remote sensing data from satellites to estimate the local sea surface slope and thus will be able to map geostrophic currents in near real time.

The frictionless nonaccelerated form of the equations of motion has also been used to derive relationships between density variations and ocean currents. For example, the "thermal wind" equation relates horizontal density gradients to vertical changes in velocity (v). In a simplified form, it is written as

$$\frac{\partial v}{\partial z} = \frac{g}{\rho f} \frac{\partial \rho}{\partial x}$$

where the x-axis is along the density gradient.

By use of this equation the strong winds of the atmosphere jet stream can be explained as the increase in wind with altitude resulting from the large density gradient that occurs across the polar front. The Gulf Stream represents an oceanographic example to which this relationship applies. The oceanic front that marks the North Wall of the Gulf Stream provides the strong density gradient that causes the rapid increase in current toward the surface characteristic of oceanic frontal regions.

A second useful relationship that can be derived from the frictionless nonaccelerated equations of motion is the "Law of Fronts,"

$$\tan \theta = - \frac{f}{g} \left[\frac{\rho_1 v_1 - \rho_2 v_2}{\rho_1 - \rho_2} \right]$$

where θ is the angle the oceanic front makes with the sea surface, ρ_1 and ρ_2 are the water densities on each side of the front, and v_1 and v_2 are the current speeds on each side of the front

This equation relates frontal slope, latitude (through the Coriolis parameter f), the density difference $\rho_1 - \rho_2$ across the front, and the velocity difference $v_1 - v_2$ across the front. Figure 2.2 is a schematic diagram of the

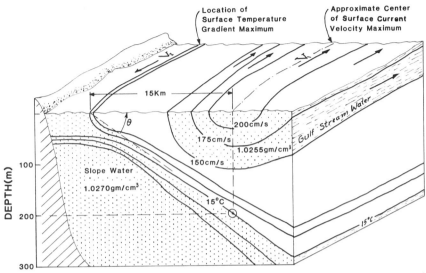

Figure 2.2. Idealized cross section of the Gulf Stream frontal zone showing frontal slope angle θ and isotherms. The maximum surface current is located approximately above the point where the 15°C isotherm intersects a depth of 200 m.

Gulf Stream front showing data from which the frontal slope can be estimated as

$$\tan \theta = - \frac{10^{-4} \ \text{rad/s}}{980 \ \text{cm/s}^2} \left[\frac{(1.0255 \ \text{g/cm}^3) \ (200 \ \text{cm/s}) - (1.0270 \text{g/cm}^3) \ (20 \ \text{cm/s})}{1.0255 \ \text{g/cm}^3 - 1.0270 \ \text{g/cm}^3} \right]$$

or

$$\tan \theta \simeq 1.26 \times 10^{-2}$$

This value for frontal slope agrees with the observed long-term average slope of the 15°C isotherm near the Gulf Stream front. This isotherm drops about 200 m in about 15 km (i.e., $\tan \theta = 1.3 \times 10^{-2}$) from its surface location on the inshore portion of the current to its offshore position at the Gulf Stream core.

2.4. FRICTION AT THE AIR–SEA INTERFACE

The nonaccelerated frictionless equations of motion have been used to develop a number of useful relations generally applicable to large-scale ocean currents. Smaller-scale turbulent exchange processes such as occur at the air–sea interface require consideration of the effects of friction. It is by this turbulent exchange of momentum that winds frictionally drag the ocean surface layers in the wind direction, producing wind-drift currents. In the following section the momentum transfer that occurs between winds and the ocean surface layers will be discussed.

2.4.1.   The Wind Factor

A common method used in a number of applied problems to estimate surface wind-driven currents is the wind factor approach, in which current speed is given as a constant percentage of the surface wind speed. The magnitude of the wind factor can be estimated if a balance between the turbulent shearing stresses on either side of the water surface is assumed. The magnitude of the shear stress is expressed as

$$\tau = \rho \, C_D \, U^2$$

where ρ is the fluid density, U is the fluid speed, and C_D is a constant of proportionality called the drag coefficient. Wind near the sea surface drags the surface ocean layer in the windward direction. This motion is transferred to subsurface layers by a turbulent-eddy exchange process. The subsurface layers are also dragged by the surface layers in the same general direction as the wind. If the wind has been blowing steadily for several hours the stress in the air, τ_a, can be equated to the stress in the water, τ, such that at the air–sea interface

$$\tau_a = \tau$$

or

$$\rho_a \, C_D \, U_a^2 = \rho \, C_D \, U^2$$

where ρ_a is the air density and ρ is the water density. The wind factor, k, is defined as the ratio of water speed to wind speed, or

$$k = \frac{U}{U_a} = \sqrt{\frac{\rho_a}{\rho}} \simeq 0.03$$

Actual measurements indicate a range in values for the wind factor from about 1 to 4 percent and an angle between surface wind direction and the surface wind-driven current of between 0 and 25°. Since wind speed usually increases rapidly with altitude in the atmosphere the value of surface wind speed U_a needs to be specified at a standard height, usually taken as 10 m or 19.5 m above the sea surface.

The wind-factor has been a useful tool for making rapid estimates of the open ocean wind-driven currents under steady conditions. In the application of the wind-factor approach, surface currents are generally estimated as 3 percent of the 10 m wind value directed at 20° to the right of the wind in the Northern Hemisphere and 20° to the left of the wind in the Southern Hemisphere. This generally agrees with observations and with a theoretical model of the wind-driven currents developed by Ekman (1905), which will be discussed in Section 2.5. The wind factor approach has been used successfully for oil spill trajectory calculations and search and rescue operations at sea.

2.4.2. A Practical Model for Transient Wind-Driven Currents

The wind factor approach is appropriate for estimates of local wind-driven currents under conditions of constant wind. However, a number of marine activities require a knowledge of transient wind-driven currents. These include coastal search and rescue operations, navigation in narrow straits, and real-time calculations of oil spill trajectories. James (1966a) developed a practical method of specifying transient wind-driven currents using a series of wind-drift forecasting curves. The curves were based on the logic that both wind waves and wind-driven currents grow as a function of wind speed, wind duration, and the distance, or fetch, over which the wind is blowing. By analogy to wind wave growth it was also assumed that a steady-state value for wind-drift would be reached for very long durations or long fetches. In this approach the deflection angle between the surface wind direction and the current was taken to be 20°. Although this angle probably varies with duration, fetch, wind speed, water depth, and other factors, the constant 20° value was considered adequate by James for rapid calculations.

Figure 2.3 shows the James wind-drift forecasting curves. The importance of this approach is that transient wind-driven current predictions can be made without detailed estimates of the air–sea momentum transfer process. To use the James curves, locate the wind speed at the top of the graph and drop vertically to the duration over which the wind has been blowing. Duration and speed can be estimated from wind observations or from available weather charts. Next read the wind-drift current at this point on the chart. Repeat this using the fetch distance instead of the duration. The lower value of the two is the wind-drift current. For example, assume a 28-knot (14.4 m/s) wind has been blowing for the past 24 hours over a fetch of 100 nautical miles (185 km). Entering values of 28 knots and 24 hours duration on Figure 2.3, current speed is found to be 0.67 knots (34.4 cm/s). Again, entering values of 28 knots and a fetch of 100 nautical miles, current speed is found to be 0.49 knots (25.2 cm/s). The smaller fetch-limited quantity is the correct value to use. If the fetch is unknown, use the value found with wind speed and wind duration alone. The curves actually give the current that would develop for a given wind starting from zero current. If, for example, wind speed is increasing, the curves must be read in a slightly different fashion.

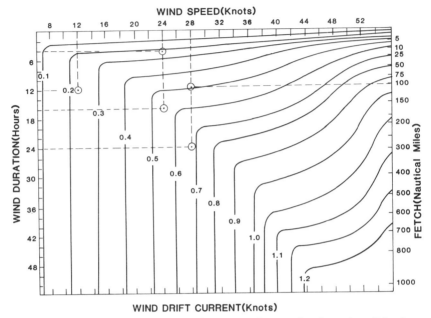

Figure 2.3.   The James wind-drift forecasting curves. See text for discussion. (After James, 1966.)

Compensation for the existing drift current is made by computing the duration that would be needed for the present wind to generate the current already present. James called this the equivalent duration. This is added to the existing duration. The procedure effectively accumulates the momentum that has already been transmitted to the current by past winds and thus permits the forecast curves to be used for cases of increasing wind. For example, if a wind blows for 12 hours at 12 knots (6.2 m/s) and then for 12 hours at 24 knots (12.4 m/s), the procedure is as follows. During the first 12 hours, the 12-knot wind generates a current of 0.22 knots (11.3 cm/s). A wind speed of 24 knots could create the same current in 4 hours, a value obtained by entering a wind speed of 24 knots on Figure 2.3 and reading 4 hours on the left axis. Adding this 4-hour duration to the 12 hours the 24-knot wind has actually blown gives an effective duration of 16 hours. Using 16 hours rather than 12 hours with the 24-knot wind speed gives the correct current speed of 0.54 knots (27.8 cm/s).

2.5.   COASTAL UPWELLING

Upwelling, the slow upward motion of deeper water that occurs in regions of surface current divergence, is important because of its tendency to enhance local nutrient and fish abundance and its relationship to fog formation at sea. Coastal upwelling is common along the western coast of continents, where favorable winds combine with the effect of the Earth's rotation to cause an offshore transport of surface waters. Upwelling then fills this region of surface current divergence. Prime coastal upwelling regions are the west coasts of the United States, North Africa, and South America. The coast of Peru is a particularly important upwelling region because of its relationship to one of the world's richest fishing grounds.

Because of its importance to the fishing industry and marine meteorology, the dynamics of coastal upwelling have long been a topic of oceanographic study. Early attempts to model upwelling were based on work by Ekman in 1905. The form of the equations of motion used by Ekman expressed a balance between the frictional drag of the surface wind and the Coriolis force. Assuming an infinitely deep, homogeneous water mass he wrote the nonaccelerated horizontal equations of motion in component form as

$$- fv = \frac{K_z}{\rho} \frac{d^2u}{dz^2}$$

$$fu = \frac{K_z}{\rho} \frac{d^2v}{dz^2}$$

The solution of this system of second-order differential equations requires two boundary conditions. At the surface Ekman assumed that the surface wind stress components were proportional to the surface current shear, or, at $z = 0$,

$$\tau_x = - K_z \frac{du}{dz}$$

$$\tau_y = - K_z \frac{dv}{dz}$$

where τ_x is the x-component of wind stress and τ_y is the y-component. The eddy viscosity coefficient K_z is the constant of proportionality. At the ocean bottom it was assumed that the wind-drift current vanished or, at $z = \infty$,

$$u = v = 0$$

The mathematical solution of these equations is called the Ekman spiral since the current components take the form of decaying spirals given as

$$u = V_0 \exp \left(- \frac{\pi}{D} z \right) \cos \left(45° - \frac{\pi}{D} z \right)$$

and

$$v = V_0 \exp \left(- \frac{\pi}{D} z \right) \sin \left(45° - \frac{\pi}{D} z \right)$$

The Ekman wind-drift current decays exponentially with depth from its surface value V_0, continuously turning to the right with depth in the Northern Hemisphere and to the left with depth in the Southern Hemisphere. The current reverses direction and is reduced to about 4 percent [exp $(-\pi)$] of its surface value at the depth of frictional influence $z = D$. The surface current is directed 45° to the right of the surface wind direction in the Northern Hemisphere and 45° to the left of the surface wind direction in the Southern Hemisphere. The surface current magnitude is calculated from the surface wind stress components using

$$V_0 = \frac{\tau_y}{\sqrt{\rho \, K_z f}}$$

where the coordinate axis is shifted so that the wind is blowing in the y-direction. The depth of frictional influence is defined as

$$D = \pi \sqrt{\frac{2K_z}{\rho f}}$$

The Ekman wind-drift current components can be integrated to the depth of frictional influence to obtain an approximation for the total wind-driven mass transport of water,

$$M_x = \rho \int_0^D u \, dz \simeq \frac{\tau_y}{f}$$

and

$$M_y = \rho \int_0^D v \, dz \simeq -\frac{\tau_x}{f}$$

Note that the two components of the Ekman mass transport are perpendicular to their respective surface wind stress components.

Figure 2.4 illustrates various aspects of the Ekman current model applied to a typical upwelling event along the California coast. Cool water is shown ascending to replace surface waters that have been transported offshore by winds blowing from the north. Upwelling season along the California coast is from April to August. During this time upwelling events occur that last for days or weeks. These episodes are marked by an increase in the northern wind component and a drop in ocean surface temperatures, causing strong thermal gradients to develop at the sea surface. Temperature-sensitive infrared satellite imagery is especially useful in providing detailed pictures of the surface location and strength of a particular upwelling event. These areas often appear as irregular light shaded regions indicating cooler temperatures

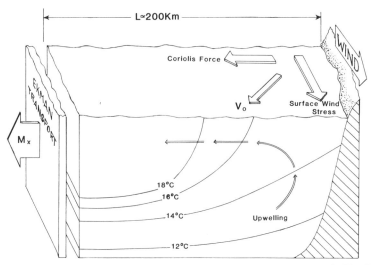

Figure 2.4. Dynamics of a coastal upwelling event as it might occur along the California coast. Pictured are the surface Ekman current V_0 directed 45° to the right of the wind and the Ekman transport M_x directed 90° to the right of the wind.

along the coastline. The pattern may also reveal details of cold-water plumes or eddies moving into the warm-water region offshore. Thermal fronts are often observed marking the offshore boundary of the upwelling region. Figure 2.5 is an enhanced infrared satellite image that shows the ocean thermal structure and frontal details during a typical upwelling event along the California coast.

Besides increasing productivity, upwelling also exerts an important influence on local meteorological conditions. For example, when upwelled water is cooler than the air just above the sea surface, it may cause this air to cool to a point at which ocean fog or low-lying stratus clouds will develop.

A simple calculation based on the Ekman model illustrates a number of dynamical relationships for coastal upwelling. During a California upwelling event the wind blows from the north typically at 1000 cm/s (about 20 knots). The surface stress for such a wind can be computed as

$$\tau_y = -\rho_a\, C_D\, U_a^2 = -(1.2 \times 10^{-3}\ \text{g/cm}^3)\,(2.0 \times 10^{-3})\,(10^3\ \text{cm/s})^2$$

or

$$\tau_y = -2.4\ \text{dyne/cm}^2$$

where ρ_a is the air density, $C_D = 2.0 \times 10^{-3}$ is a reasonable value for the drag coefficient, and U_a is the surface wind speed. The negative sign indicates wind stress in the negative y-direction (south). Available observations

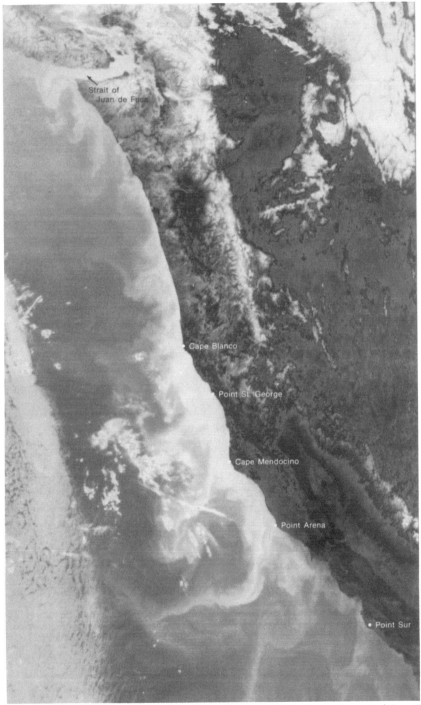

Figure 2.5. Satellite image of a California upwelling event showing cooler water along the coast and oceanic thermal fronts. (Courtesy of National Oceanic and Atmospheric Administration.)

indicate that California upwelling events influence water movements to depths of 50 to 300 m. A depth of frictional influence of 50 m leads to an estimate for the eddy viscosity coefficient K_z using

$$D = \pi \sqrt{\frac{2 K_z}{\rho f}}$$

as

$$K_z = \frac{\rho f D^2}{2\pi^2} = \frac{(1.02 \text{ g/cm}^3) (10^{-4} \text{ rad/s}) (5 \times 10^3 \text{ cm})^2}{19.7} = 129 \text{ g/cm·s}$$

The magnitude of the surface current can now be estimated as

$$V_0 = \frac{|\tau_y|}{\sqrt{\rho f K_z}} = \frac{2.4 \text{ dyne/cm}^2}{\sqrt{(1.02 \text{ g/cm}^3) (10^{-4} \text{ rad/s}) (129 \text{ g/cm·s})}} = 21 \text{ cm/s}$$

The Ekman model predicts a surface current with a magnitude of 21 cm/s directed at an angle of about 45° to the right of the wind (toward the southwest) during a typical California upwelling event. The Ekman mass transport directed 90° to the right of the surface wind, is offshore (toward the west), and can be approximated as

$$M_x = \frac{\tau_y}{f} = \frac{- 2.4 \text{ dyne/cm}^2}{10^{-4} \text{ rad/s}} = - 2.4 \times 10^4 \text{ g/cm·s}$$

Offshore transport above the depth of frictional influence tends to move the surface waters away from the California coast across an upwelling region that usually extends 100 to 200 km offshore. Cool water slowly rises to fill the divergence zone. The magnitude of this upwelling velocity may also be approximated, since it depends on the divergence of the water transported away from the coastline. Using a form of the continuity equation leads to an estimate for the upwelling velocity w_D, given in terms of the transport divergence near the coast, as

$$w_D = \frac{\tau_y}{\rho f L} = \frac{- 2.4 \text{ dyne/cm}^2}{(1.02 \text{ g/cm}^3) (10^{-4} \text{ rad/s}) (2 \times 10^7 \text{ cm})}$$

or $w_D = - 1.2 \times 10^{-3}$ cm/s at the depth of frictional influence. The divergence of the mass transport τ_y/fL is easily calculated, since the coastal value for M_x is 0 and the length of the upwelling zone offshore is $L = 200$ km. This estimate for w_D agrees with upwelling velocities that have been inferred from oceanographic observations in this region. For a more detailed treatment of the physical, chemical, and ecological implications of coastal upwelling see Richards (1981).

2.6. SEASONAL COASTAL CURRENTS

The theoretical description of currents along the coastlines of the world's oceans has been a topic of active investigation by physical oceanographers. Observations indicate that currents in these coastal oceans, defined here as regions with a water depth of less than 200 m, are composed of transient wind-driven and tidal flows superimposed on a relatively weak mean seasonal circulation pattern. The permanent seasonal component is especially important because of its influence on long-term pollutant fate and its effect on commercial fisheries.

Seasonal coastal currents can be calculated by using an extension of the Ekman model that includes the effects of bottom topography and coastal pressure gradients. Such models have been applied successfully to the New York Bight, a coastal ocean along the northeastern coast of the United States, by Csanady (1976), Bishop and Overland (1977), and Bishop (1980c). In the Bishop and Overland model the governing equations describe a balance between friction, Coriolis, and pressure gradient forces. The equations of motion take the form:

$$- fv = \frac{K_z}{\bar{\rho}} \frac{\partial^2 u}{\partial z^2} - \frac{1}{\bar{\rho}} \frac{\partial p}{\partial x}$$

and

$$fu = \frac{K_z}{\bar{\rho}} \frac{\partial^2 v}{\partial z^2}$$

where $\bar{\rho}$ is the average density from the water surface to the bottom, the x-axis points offshore perpendicular to the coastline, and the y-axis is along the coast.

The cross-shelf pressure gradient has two components:

$$\frac{1}{\bar{\rho}} \frac{\partial p}{\partial x} = g \frac{\partial \zeta}{\partial x} - \frac{gz}{\bar{\rho}} \frac{\partial \bar{\rho}}{\partial x}$$

The first component represents the permanent cross-shelf sea surface slope $\partial \zeta / \partial x$ while the second component arises from the seasonal cross-shelf density gradient $\partial \bar{\rho} / \partial x$. In addition to the cross-shelf pressure gradient some models also include estimated values for the permanent longshore pressure gradient in the analysis of seasonal coastal currents.

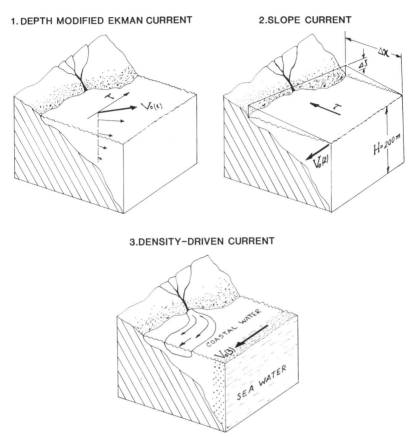

Figure 2.6. Three components of the Bishop and Overland (1977) diagnostic coastal current model.

The solution to this form of the equations of motion, given by Bishop and Overland, has three components which are illustrated in Figure 2.6. The first component is a shallow-water version of the Ekman wind-drift current driven by the climatological mean wind stress. The effect of shallow water is to cause the surface current to flow at an angle less than 45° from the wind as predicted by the Ekman deep water model. The second component in the model is a slope current which results from the mean climatological wind blowing toward or away from the coastline. The resulting cross-shelf sea surface slope drives a permanent current component which flows approximately parallel to the coast. The third component in the model results from the seasonal cross-shelf density gradient that commonly occurs in coastal waters due to the influx of fresh, less dense runoff waters near the coast.

Given the climatological values for the mean seasonal wind stress and cross-shelf density gradient, the seasonal permanent current can be cal-

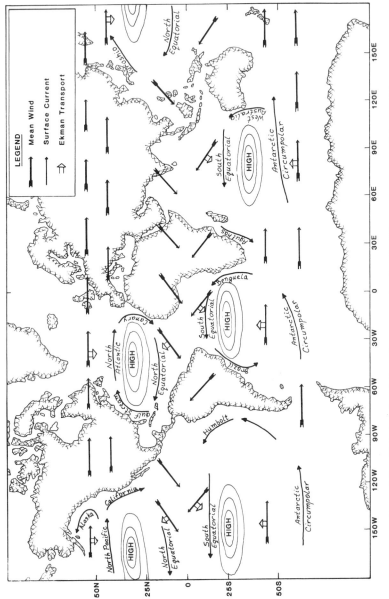

Figure 2.7. Global-scale winds and currents, showing the Ekman transport and the resulting increase in mid-ocean sea surface elevation. The general circulation of the oceans may be considered a response to the pressure gradient induced by this mid-ocean rise in sea surface elevation.

culated from the solution to the model equations. Since this approach is actually a combination of theory and observation, the calculation is called a diagnostic model. Diagnostic models have proven to be an important complementary source of information on currents, especially for coastal oceans.

2.7. GENERAL CIRCULATION OF THE OCEANS

The mean winds provide the driving mechanism for the general circulation of the oceans on a global space scale and a seasonal time scale. Although global oceanic flow represents a complicated geophysical process, a first-order explanation can be developed based on Ekman's model of wind-driven currents. Since the Ekman transport is 90° to the right of the wind in the Northern Hemisphere and 90° to the left in the Southern Hemisphere, the alternating westerly and easterly winds in each hemisphere produce oceanic convergence zones at about 30° latitude in each ocean basin. The converging water actually produces a "mound" in sea surface topography on the order of meters in height when measured across the whole basin. These regions of higher sea surface elevation are responsible for the global-scale pressure gradients that drive large-scale circulation gyres. For example, the North Atlantic gyre is composed of the Gulf Stream, the North Atlantic Current, the Canary Current, and the North Equatorial Current, as shown in Figure 2.7. The figure also indicates how the circulation around the gyres in the Southern Hemisphere is in the opposite direction to that in the Northern Hemisphere. In all gyres water initially flows down the pressure gradient from higher toward lower levels. As it moves it is turned by the Coriolis force until it has reached approximate geostrophic equilibrium (see Figure 2.1). The result is that global-scale currents of each gyre flow approximately parallel to the lines of constant pressure (elevation) in the sea surface topography.

The western boundary currents of all the ocean basins, such as the Gulf Stream in the North Atlantic basin, flow about 10 times faster than eastern boundary currents such as the Canary Current. This westward intensification of the global circulation cannot be explained by the simple Ekman transport model; this requires more sophisticated models that include the effects of the ocean boundary and the variation of Coriolis force with latitude. A review of these global-scale wind-driven ocean circulation models can be found in Kamenkovich (1977).

Three

OCEAN WIND WAVES

Specification of the wind-generated sea state is important for a number of ocean use activities such as marine transportation and marine resource management. In addition to wind waves, the main subject of this chapter, the full spectrum of ocean waves also includes the astronomical tides, storm surges, and internal waves, all to be discussed in Chapter 4. Wind waves occupy that portion of the ocean wave spectrum covering waves with periods of about 30 to 0.1 s. Within this spectral band the wind-generated sea state can be described by a spectrum of simple harmonic component waves, each with a specific wave period. This spectral description of wind waves is the cornerstone of modern wave prediction and has direct applications in shiprouting, offshore structure design, and naval operations at sea.

3.1. THE HARMONIC COMPONENT WAVE

The ocean surface can be described as a superposition of simple wave forms. A formal mathematical representation of ocean surface waves is based on the harmonic component wave

$$\eta(x,t) = a_n \cos\left(\frac{2\pi}{L} x - \frac{2\pi}{T} t\right)$$

were $\eta(x,t)$ is the harmonic wave form moving in distance x and time t, with amplitude a_n, wavelength L, and period of oscillation T. Additional simple harmonic wave parameters are the wave number $k = 2\pi/L$, the angular frequency $\omega = 2\pi/T$, and the frequency $f = 1/T$. Figure 3.1 shows a harmonic wave progressing along the x-axis at time $t = 0$ with a phase speed

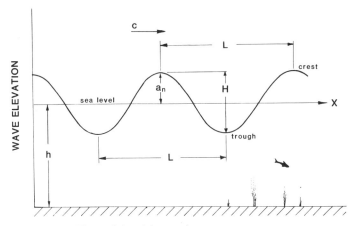

Figure 3.1. A harmonic component wave.

$c = L/T = \omega/k$. The phase speed of the harmonic component may also be expressed in terms of water depth h as

$$c = \sqrt{\frac{g}{k} \cdot \tanh kh}$$

where g is gravity and tanh is the hyperbolic tangent. This equation has limiting values for both deep and shallow water, defined in terms of the ratio of water depth to wavelength. If h/L is large the water is considered to be deep. In this case tanh $kh = 1$ and $c = \sqrt{g/k} = \sqrt{gL/2\pi}$. Using $c = L/T$ leads to the relationship between deep water phase speed and wave period $c = (g/2\pi)\,T$. If h/L is small, the water is considered to be shallow, tanh $kh = kh$ and $c = \sqrt{gh}$. Thus, in shallow water phase speed depends on water depth, while in deep water phase speed depends on wave period or wavelength. Water is considered deep for h/L values greater than $\frac{1}{2}$ and shallow for h/L values less than $\frac{1}{20}$. For wind waves on the open ocean h/L is large, so harmonic component waves propagate at speeds that depend on their wavelengths and wave periods. This is illustrated by the commonly observed dispersion phenomenon in which the longest waves move the fastest away from the wave generation area. On the other hand, tidal waves, which are generated by the gravitational attraction of the sun and moon, are very long waves. Since the ratio of water depth to wavelength for these waves is always small, the water is always considered shallow and phase speed is a function of water depth.

The average wave energy of a harmonic component wave per unit surface area is proportional to the square of the wave amplitude a_n and is given by

$$E_n = \tfrac{1}{2}\,\rho g\,a_n^2$$

where ρ is the water density. The speed at which this wave energy is transported over the ocean surface is called the group speed and is given by $c_g = \frac{1}{2}c$ for deep water waves and by $c_g = c$ for shallow water waves. Under certain conditions standing waves, or seiches, can form in enclosed bodies of water. This type of wave oscillates with a period that depends on the dimensions of the basin.

3.2.   THE OCEAN WAVE SPECTRUM

The ocean surface is a random collection of waves of various shapes and sizes. A formal mathematical representation of this surface uses the harmonic component wave as a building block and expresses the sea surface as a combination of a finite number of these component waves, each with its own amplitude, frequency, and direction. Wave records representing a series of wave observations at a given point show a wide range of amplitudes, periods, and propagation directions. Harmonic analysis, also called time series analysis or Fourier analysis, is frequently applied to such data to extract meaningful information. This is done by decomposing the time series into harmonic components, each having a particular frequency and amplitude. Harmonic analysis is based on the relationship between a time series, in this case a wave record, and the sum of harmonic component waves

$$\eta\,(t) = a_0 + \sum_{n=1}^{r} a_n \cos\,(n\omega_n t + \epsilon)$$

where $\eta(t)$ = time series of wave heights at a point in the ocean
$\quad\ \ a_0$ = mean wave height over the wave record
$\quad\ \ \omega_n$ = angular frequency of one harmonic component
$\quad\ \ n$ = identifying number of the harmonic component
$\quad\ \ a_n$ = amplitude of the nth harmonic component
$\quad\ \ r$ = total number of harmonic components used in the analysis
$\quad\ \ \epsilon$ = phase angle of the nth component

Figure 3.2 illustrates how harmonic analysis may be applied to a hypothetical ocean wave record taken at an observation site. Since wave energy is proportional to wave amplitude squared, the results of the time series analysis (wave amplitudes and frequencies) are generally expressed as an energy spectrum. A spectrum is a plot of spectral density $S(\omega_n)$, which is proportional to wave energy, and frequency. The relationship between spectral density and wave amplitude is

$$\tfrac{1}{2}\,a_n^2 = \lim_{\Delta\omega\to 0}\,[S\,(\omega_n)\Delta\omega]$$

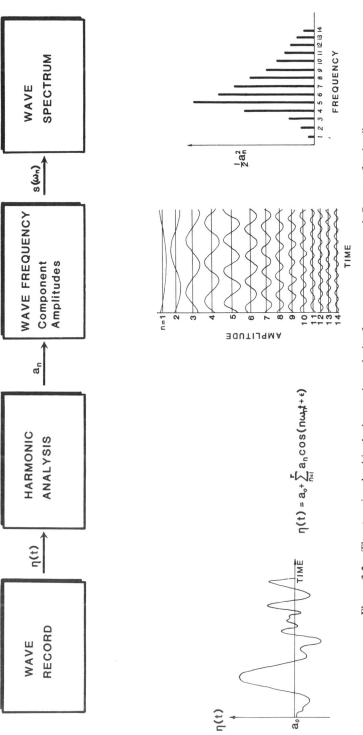

Figure 3.2. The steps involved in the harmonic analysis of an ocean wave record. See text for details.

$$\eta(t) = a_o + \sum_{n=1}^{r} a_n \cos(n\omega_1 t + \epsilon)$$

47

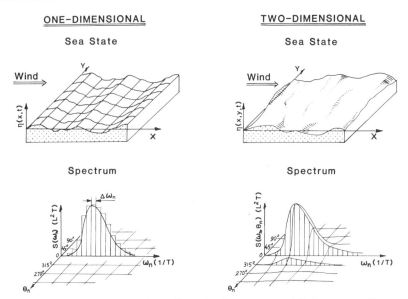

Figure 3.3.   One-dimensional and two-dimensional sea states and their corresponding energy density spectra.

where $\Delta\omega$ is the angular frequency bandwidth between spectral components. The spectral density function can be expressed in terms of frequency rather than angular frequency by using the conversion relationship $S(f_n) = 2\pi S(\omega_n)$. Spectral density has dimensions L^2T.

To understand the properties of the ocean wave spectrum, consider an idealized ocean wave field composed of a number of one-dimensional or long-crested waves all propagating in one direction, changing shape as they move. Harmonic analysis of this wave motion produces a frequency spectrum composed of a number of components each having some wave amplitude a_n and some angular frequency ω_n. Figure 3.3 shows the relationship between this type of wave motion and the resulting energy density spectrum.

Under actual conditions at sea, waves are not long-crested and do not all travel in the same direction. For such two-dimensional or short-crested waves a more complicated time series analysis that accounts for both time and space variations of wave elevations is required. The resulting directional spectrum distributes energy density across both frequency and direction of propagation. Figure 3.3 illustrates a directional spectrum $S(\omega_n,\theta_n)$ that would result from a wind blowing along the x-axis. It shows the range of harmonic components produced at various directions θ_n. Generally the local sea state consists of local wind-generated waves plus low-frequency swell generated by distant storms. Accordingly, the directional spectrum may have two peaks, one representing local sea conditions, such as that illustrated in Figure 3.3, and a second, sharper peak representing incoming swell.

The total spectral energy E in a given sea state is calculated by summing the contributions from each spectral component across all frequencies and directions. Expressed mathematically, the total spectral energy is

$$E = \int_{0°}^{360°} \int_0^\infty S(\omega_n, \theta_n) \, d\omega d\theta$$

It should be noted that total spectral energy has the dimensions of length squared (L^2), because the constant ρg used in the definition of component energy E_n is usually omitted.

Various statistical parameters can be calculated from E. The most widely used parameter, significant wave height ($H_{1/3}$), is given as

$$H_{1/3} = 4.0\sqrt{E}$$

Significant wave height is defined as the average height of the highest one-third waves observed at a point and may be calculated from a harmonic analysis of a wave record. Significant wave height is a particularly useful parameter because it is approximately equal to the wave height that a trained observer would visually estimate for a given sea state.

Idealized one-dimensional frequency spectra have been developed based on theory and careful analysis of winds and measured wave heights. One of the most widely used spectral forms is the Pierson–Moskowitz spectrum for fully developed seas (Pierson and Moskowitz, 1964). A fully developed sea is one in which each spectral component has reached its maximum amplitude for a given wind speed. At this point, energy input from the wind is balanced by energy loss by dissipative processes such as wave breaking. The Pierson–Moskowitz spectrum is expressed as

$$S_\infty(\omega) = \frac{\alpha g^2}{\omega^5} \exp\left[-\beta \left(\frac{\omega_0}{\omega} \right)^4 \right]$$

where $S_\infty(\omega)$ = a maximum value (given in $cm^2 s$) for spectral density at angular frequency ω, given a specific wind blowing over a sufficiently long duration and fetch

α = 8.1×10^{-3}
g = 980 cm/s^2
β = 0.74
ω_0 = $g/U_{19.5}$ (a constant for a given wind speed)
$U_{19.5}$ = wind speed in cm/s measured at 19.5 m above sea level

The constants α, β, and ω_0 were empirically determined from wind and wave observations taken in the North Atlantic Ocean.

The Pierson–Moskowitz spectral form has been used for calculating the fully developed spectral energy for a given wind speed:

$$E_\infty = \int_0^\infty S_\infty(\omega)\, d\omega = \frac{\alpha}{4\beta g^2}\, U_{19.5}^4$$

and the fully developed significant wave height for a given wind speed:

$$(H_{\frac{1}{3}})_\infty = 4.0\,\sqrt{E_\infty} = 2.12 \times 10^{-2}\, U_{19.5}^2$$

where $(H_{\frac{1}{3}})_\infty$ is the maximum significant wave height in meters and $U_{19.5}$ is wind speed in m/s. The peak of the Pierson–Moskowitz spectrum is at the point where

$$\frac{\partial S_\infty(\omega)}{\partial \omega} = 0$$

which leads to

$$\omega_p = \left(\frac{4\beta}{5} \right)^{1/4}\omega_0 = 0.877\,\omega_0$$

where ω_p is the angular frequency at which maximum wave energy occurs.

3.3. SPECTRAL GROWTH THEORIES

One extension of the spectral representation for ocean wind waves has been the development of theories of the growth of a spectral component under the action of a constant wind. A simple model can be used to calculate the initial stages of component growth. For the case of unlimited fetch the growth rate of spectral density S is expressed as

$$\frac{\partial S}{\partial t} = A + BS$$

where $\partial S/\partial t$ is the local rate of change in spectral density. The first growth term, A, represents a linear growth, since $\partial S/\partial t = A$ implies growth of the form $S = S_0 + At$ where $S = S_0$ at time $t = 0$. B is an exponential growth term, since $\partial S/\partial t = BS$ implies growth of the form $S = S_0 \exp(Bt)$ where, again, $S = S_0$ at time $t = 0$. Each of these terms is, in general, a function of frequency, direction of component propagation, position, time, and wind velocity.

The linear growth term is derived from a theory developed by Phillips (1957), while the exponential growth term is derived from a theory first developed by Miles (1957). In the growth equation given above, called the Miles–Phillips model, initial linear growth results from a resonance between traveling atmospheric pressure fluctuations and wave components of similar dimensions. This mechanism is effective in initiating growth from a calm sea state. The exponential growth mechanism operates on waves already in existence by supplying energy from the mean wind directly to the waves. The combined Miles–Phillips component growth model predicts an initial linear growth phase governed by the resonance process, followed quickly by an exponential growth phase. This model does not describe any mechanism for stopping or slowing growth once it has been initiated.

In addition to the limitations resulting from fetch and wind duration, various nonlinear processes, such as wave breaking and the energy transfer between components, tend to limit growth. Assuming that the Pierson–Moskowitz fully developed spectral value is the appropriate limit for each component, Inoue (1967) developed a modified Miles–Phillips growth model using S_∞ as a mathematical cutoff such that

$$\frac{\partial S}{\partial t} = \left[A \left\{ 1 - \left(\frac{S}{S_\infty} \right)^2 \right\}^{\frac{1}{2}} + BS \right] \left[1 - \left(\frac{S}{S_\infty} \right)^2 \right]$$

The growth curve represented by this equation exhibits an initial linear phase, then an exponential phase, and finally an equilibrium phase approaching the limiting value of the Pierson–Moskowitz spectrum, S_∞.

Although the Inoue model has proved very useful for wave forecasting applications, some observations (see, for example, Barnett and Wilkerson, 1967) have indicated that an additional growth phase, namely an overshoot, should be added to component growth models. Figure 3.4 shows schematic

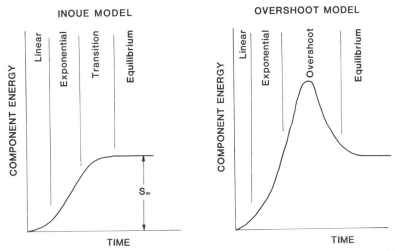

Figure 3.4. Schematic growth curves for component energy according to the Inoue model and the overshoot model. (From Bishop, 1974.)

growth curves for the Inoue and overshoot models. Component energy in the overshoot model reaches a value 120 to 200 percent higher than its final equilibrium value. The overshoot feature of spectral component growth has led some investigators to conclude that a nonlinear transfer of energy between spectral components might be an important factor in the growth process especially in the final stages.

3.4. WIND FIELD ANALYSIS TECHNIQUES

The single most important factor in the prediction of waves at sea is the correct specification of the wind field. Wind observations reported by ships are considered only a secondary data source for the development of marine wind fields because of the sparseness of the data. Instead, wind data used for wave calculations are derived from the surface atmospheric pressure patterns given on marine weather charts. These charts are generally available at 6-hour intervals and are used to establish wind velocity, fetch, and duration. This information, combined with the geostrophic equation, forms the basis for the development of wind field data used in wave calculations.

The first step in determining wind fields from surface weather charts is to specify a fetch. Fetch is the extent of open ocean over which the wind blows to produce waves and refers to an oceanic region over which wind speed and direction are reasonably constant. Fetch lengths can be identified on marine weather charts as areas where isobars are parallel. Fetch boundaries occur at coastlines, where isobars start to curve, spread, or converge, and at meteorological fronts where wind shifts occur. Estimates of wind duration can be obtained by comparing wind fields from consecutive charts. The minimum duration that can be established from marine weather charts is 6 hours.

Surface wind speeds are calculated from atmospheric pressure gradients estimated from marine weather charts using the geostrophic equation

$$V_g = \frac{1}{\rho_a f} \frac{\Delta p}{\Delta x}$$

where V_g is the geostrophic wind speed, ρ_a is the density of air, f is the Coriolis parameter, and $\Delta p / \Delta x$ is the horizontal gradient of surface atmospheric pressure as derived from a marine weather chart.

Friction near the sea surface reduces the geostrophic wind and causes the wind to cross isobars toward low pressure at an average angle of 10 to 20°. The stability of the air also affects the wind. A stable condition occurs when the air mass is warmer than the sea surface, whereas an unstable condition occurs when the air is colder than the sea surface. Winds at sea are stronger in unstable than in stable conditions, given the same pressure gradient. Table 3.1 shows a commonly used air mass stability correction factor that

TABLE 3.1.   STABILITY CORRECTION FOR GEOSTROPHIC WINDS

Sea Temperature minus Air Temperature (°F)	Sea Temperature minus Air Temperature (°C)	Ratio of Surface Wind Speed to Geostrophic Wind Speed, V/V_g
negative	negative	0.60
0 to 10	0 to 5.6	0.65
10 to 20	5.6 to 11.2	0.75
above 20	above 11.2	0.90

Source: U.S. Army, *Shore Protection Manual* (1973).

can be applied to calculated geostrophic winds at sea. A detailed analysis of the effects of atmospheric stability on ocean surface winds can be found in Cardone (1969).

For an example of how marine weather charts are used to develop wind field data for wave prediction, consider Figure 3.5. This chart of the North Pacific Ocean shows two low-pressure systems. One storm is located at latitude 44°N and longitude 129°W, and a second storm is located at latitude 53°N and longitude 146°W. A high-pressure region is also shown centered at approximately latitude 34°N and longitude 152°W. Three fetch regions (1,2, and 3) are shown on the chart as shaded areas. For each region wave prediction sites and the respective fetch lengths F_1, F_2, and F_3, are also shown. The geostrophic wind equation can be used to estimate the surface winds. In region 1, for example, the wind results from a pressure gradient of about 18 millibars over about 740 km, giving

$$V_g = \frac{18 \times 10^3 \text{ dyne/cm}^2}{(1.3 \times 10^{-3} \text{ g/cm}^3)(10^{-4} \text{ rad/s})(740 \times 10^5 \text{ cm})}$$

$$V_g = 19 \text{ m/s } (\sim 37 \text{ knots})$$

In this region the air behind the cold front is at least 20°F (11.2°C) colder than the sea water. Applying the sea–air stability correction (Table 3.1) gives $V = 0.9V_g \simeq 17$ m/s (~33 knots) as the surface wind magnitude. The wind direction in region 1, as indicated from the isobar orientation, is from the northwest, or from about 315° True. Friction causes the actual surface wind direction to be about 15° towards low pressure, or from about 300° True. In regions 2 and 3 the pressure gradients are about 50 percent greater than in region 1, so that the geostrophic winds are about 50 percent stronger. From this example it is clear that a careful analysis of the surface wind field, a critical factor for ocean wave predictions, must be developed from the rather limited information given on marine weather charts.

Wind data used for input into wave models can also be obtained from computer forecasts of the marine surface pressure field or from idealized

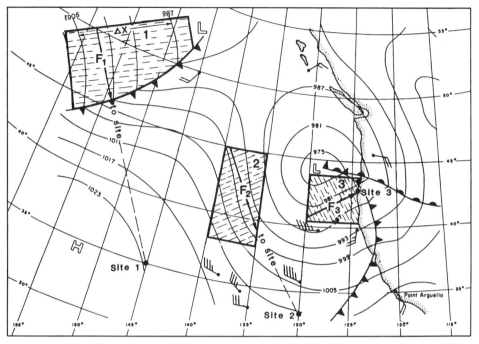

Figure 3.5.   Surface weather analysis chart showing pressure in millibars and ship winds reported as barbed arrows (full barb = 10 knots, half barb = 5 knots). Three fetch regions, 1, 2, and 3, three fetch lengths, F_1, F_2, and F_3, and three wave prediction sites are also shown. (After U.S. Army *Shore Protection Manual*, 1973.)

analytical equations. Computer models of marine weather conditions are used by various countries throughout the world. In the United States the National Weather Service and the Navy Fleet Numerical Oceanography Center produce computer-analyzed marine pressure charts that can be used as input for wave calculations. Wind data used for wave forecasts in specific tropical cyclones are generally derived from parametric storm models. These are simply mathematical equations that represent the change of wind radially away from the storm center. The key parameters in these models are the storm's central pressure, radius, and forward motion. A complete review of computer and parametric wind models used as input for ocean wave predictions has been prepared by Overland (1979).

3.5.   WAVE PREDICTION MODELING

Wave modeling has evolved through the years. Manual methods, based on empirically derived graphs and tables, were used effectively to give site-specific predictions as an aid to military operations during World War II. After the war, global computer-based forecasting models evolved to meet developing military needs. Applications of wave prediction models now include deep-water forecasts for commercial and military shiprouting, nearshore forecasts for commercial and recreational interests, and climatological forecasts of extreme wave conditions for ocean engineering applications such as offshore structure design.

Wave models are characterized by the types of output information they provide. Significant-wave techniques provide significant wave heights and periods calculated from empirical relationships between the wind and wave parameters. Spectral techniques provide wave energy as a function of frequency and direction. Discrete spectral models numerically calculate wave energy spectra, while parametric spectral models specify wave energy by spectral shape parameters and directional spreading functions. Hybrid spectral models combine aspects of both discrete and parametric spectral wave models.

Manual and computer solutions are employed for both significant-wave and spectral prediction techniques. Examples of the manual approach are the significant-height method, first developed by Sverdrup and Munk (1947), the spectral method, developed by Pierson, Neumann, and James (1955), and the parametric method, used by Ross (1976). Examples of the computer approach include the numerical significant-wave-height method developed by Wilson (1961, 1965), the energy-balance equation method used by Pierson et al. (1966) and Cardone et al. (1976), and the hybrid parametric method developed by Gunther et al. (1979). A brief outline is given below of these representative wave prediction approaches (listed in Table 3.2) to illustrate how wave models have been applied to the calculation of waves.

The Sverdrup-Munk method was first developed for site-specific forecasts during World War II amphibious operations. As improved by Bretschneider (1952, 1970) and referred to as the SMB method, it is the basis for manual significant-wave calculations. Given wind speed, direction, duration, and fetch, which may be derived from available marine weather charts, significant wave heights and periods are estimated from empirically derived wave prediction nomographs. Although the approach proved very useful for plan-

TABLE 3.2. WAVE MODELS CATEGORIZED BY OUTPUT AND METHOD OF SOLUTION

Method of Solution	Output		
	Significant Wave	Spectral	
		Discrete	Parametric
Manual	SMB	PNJ	Ross
Computer	Wilson	Pierson	Gunther

ning World War II military operations, the SMB method has had only limited use in present day applications, since it does not provide the spectral distribution of the various wave components. The SMB method is described in detail in the U.S. Army *Shore Protection Manual* (1973).

The Pierson, Neumann, and James (1955) manual wave prediction technique (the PNJ method) uses an integrated version of the wave spectrum called the co-cumulative spectrum (CCS). Figure 3.6 illustrates the relationship between the amplitude spectrum, $2S(f_i)$, and the co-cumulative spectrum. Using the CCS the forecaster reads off the total energy E^* based on wind speed, duration, and fetch, which are derived from an analysis of available marine pressure charts. As with the James forecasting curves for wind-driven currents discussed in Chapter 2, the wave height may be either duration- or fetch-limited for a specific wind speed. Once the value of E^* is found on the CCS curve, the significant wave height can be quickly calculated. This is shown in Figure 3.7. The principal spectral frequencies (or periods) in the sea state are those to the right of the intersection of the CCS and the appropriate fetch or duration curves. In Figure 3.7 the principal wave periods are therefore less than 11 s.

Ross (1976) used a parametric manual method based on the theoretical work of Hasselmann et al. (1976) to forecast waves. The technique, devel-

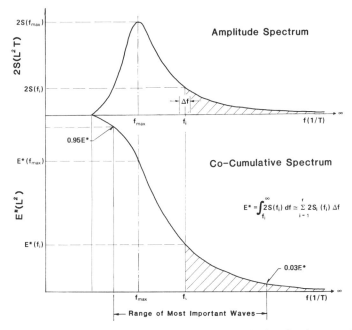

Figure 3.6. Relationship between the amplitude spectrum, which is twice the spectral density, $[2S(f_i)]$ and the total energy E^* of the co-cumulative spectrum (CCS) as used in the PNJ method.

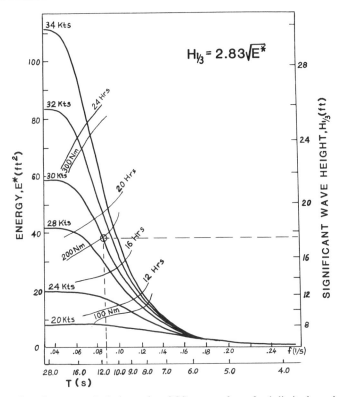

Figure 3.7. Sample wave calculation using CCS curves for a fetch-limited sea in a 30-knot (15.4 m/s) wind blowing for 24 hours over a 200-nautical-mile (370 km) fetch giving $H_{1/3} \simeq 17$ ft (5.2 m) and $T \simeq 11$ s. In the PNJ method total energy E^* is defined such that significant wave height $H_{1/3}$ is given as $2.84\sqrt{E^*}$.

oped specifically for tropical cyclones, employs nondimensional plots. The key input parameters are wind speed U and radial distance R from the storm center. Output parameters are significant wave height (or total spectral energy) and the frequency of the spectral peak. The simple plots allow rapid manual calculations to be made during actual storm conditions, as shown in Figure 3.8. Long (1979) gives an example of a calculation using this parametric manual method for Hurricane Eloise, a 1975 tropical cyclone that occurred in the Gulf of Mexico. The calculation was done for a data buoy site where winds measured 38 m/s. The buoy was 29 km from the storm's center. Maximum significant waves observed at the buoy were about 9 m in height. The first parameter needed in the calculation is the dimensionless distance from the storm center, given as

$$\frac{Rg}{U^2} = \frac{(29{,}000 \text{ m}) (9.8 \text{ m/s}^2)}{(38 \text{ m/s})^2} = 197$$

where g is gravity. Using Figure 3.8, the dimensionless peak frequency is read as 0.32 and the dimensionless significant wave height is 0.062. The peak frequency can now be found from

$$\frac{f_m U}{g} = 0.32$$

$$f_m = \left(\frac{9.8 \text{ m/s}^2}{38 \text{ m/s}} \right) (0.32) = 0.083 \text{ 1/s}$$

and the significant wave height from

$$\frac{H_{1/3} g}{U^2} = 0.062$$

$$H_{1/3} = \left[\frac{(38 \text{ m/s})^2}{9.8 \text{ m/s}^2} \right] (0.062) = 9.1 \text{ m}$$

These values agree with the maximum significant wave height and peak frequency actually observed at the buoy during the storm. The advantage of the manual parametric approach is that the total spectral energy (or signifi-

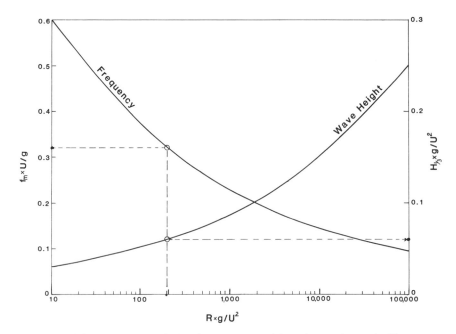

Figure 3.8. The Ross parametric hurricane wave model used to estimate significant wave height $H_{1/3}$ and peak spectral frequency f_m, given location R and wind speed U in the storm. (After Long, 1979.)

cant wave height) and peak frequency can be estimated rapidly from a limited number of input parameters. A disadvantage is that it does not allow direct calculation of the full directional spectrum. In addition, some investigators feel that this approach might not give accurate results for fast-moving storms in which local seas have not adjusted to the fetch and wind speed parameters because of the short duration of the wind.

Wilson (1961, 1965) developed a computerized version of the SMB manual method applicable to moving storms in which the wind changes over time and space. He derived a mathematical function, based on the original SMB curves, for small time and fetch increments. The Wilson model computes wave growth, propagation, and decay for fixed forecast sites. Surface weather charts provide input for the model. The model output is a prediction over time of significant wave height and period at the forecast sites. For many practical applications this approach represents an important improvement over the SMB manual method, because it considers wind fields that change in time and space, it automatically accounts for wave decay, and it is easily programmed on a computer. A disadvantage is that being a significant-wave model, it does not provide the frequency or directional spectrum necessary for many applications.

The most widely used type of computerized wave modeling technique is based on the energy-balance equation. In this approach an advanced form of the Miles–Phillips growth equation is used to predict the two-dimensional wave frequency spectrum as it grows, propagates, and decays. The basis of this method is an equation that "balances" energy input by wind with energy removed by wave breaking and dissipation and that may include the nonlinear transfer of energy between spectral components. The balance equation is

Local Rate of Change in Spectral Energy	=	Energy Transport at Group Velocity	+	Effects of Depth and Currents	+	Sources and Sinks of Component Energy

▲

Although the source and sink function, the biggest unknown in the equation, represents an area of ongoing research, it is generally assumed to include the following parts:

$$SF = SF_{in} + SF_{out} + SF_{nl}$$

where SF is the net source or sink function for a single spectral component, SF_{in} represents sources of input energy to a spectral component and is usually taken from the Miles–Phillips theory, SF_{out} is the component energy removal (sink) term representing the limiting effects of the wave breaking process, and SF_{nl} represents weak nonlinear processes which transfer energy between spectral components (a source or sink, depending on the component's location within the spectrum).

The computerized spectral wave model of Barnett (1968) uses a form of energy-balance equation which includes the transport of component energy across the ocean, a modified Miles–Phillips component growth mechanism, a dissipation function that limits component growth due to wave breaking, and a nonlinear transfer term that accounts for the movement of spectral energy between components. In terms of spectral energy density, S, the energy equation is

$$
\begin{matrix}
\text{Local} \\ \text{Change}
\end{matrix}
=
\begin{matrix}
\text{Energy} \\ \text{Transport} \\ \text{Term}
\end{matrix}
+
\begin{pmatrix}
\text{Miles–Phillips} \\ \text{Growth} \\ \text{Term}
\end{pmatrix}
\cdot
\begin{pmatrix}
\text{Wave} \\ \text{Breaking} \\ \text{Term}
\end{pmatrix}
+
\begin{matrix}
\text{Nonlinear} \\ \text{Transfer} \\ \text{Terms}
\end{matrix}
$$

$$
\frac{\partial S}{\partial t} = - C_g \frac{\partial S}{\partial x} \qquad + \qquad (A + B\ S)\cdot(1 - \mu) \qquad + (\Gamma - \tau_0)
$$

where C_g is the component's group velocity, μ is a cutoff function that simulates wave breaking, and Γ and τ_0 are nonlinear energy transfer functions. The Barnett model was used to solve the balance equation at grid points covering the North Atlantic Ocean with a separation of 120 nautical miles and with a time step of 1 hour. Using marine surface wind fields as input, the model provided a time series of directional wave spectra at each grid point that compared well with the available observations.

The computerized spectral model represents the most detailed attempt to predict the global ocean wave field on a real-time basis. The Navy Fleet Numerical Oceanography Center (FNOC) uses this type of model for global operational wave predictions. The FNOC model was first developed by Pierson et al. (1966). Later a modified version was used by Cardone et al. (1976) to calculate hurricane waves in the Gulf of Mexico. Figure 3.9 shows an example of the output from this model for Hurricane Eloise (the same storm used in our example of the Ross parametric model). Good agreement between predicted and observed wave heights and frequencies was obtained using this model, with the additional advantage over the parametric model of giving the spectral distribution of wave heights and frequencies with direction. The disadvantage of computerized spectral models is their high operating cost.

The hybrid parametric method combines aspects of a parametric model with the wave propagation aspect of the balance equation model. In the parametric model the local wave frequency spectrum is determined using key parameters such as wind speed and fetch. In the hybrid parametric model, frequency spectrum values are multiplied by a directional spreading function to produce local two-dimensional energy spectra. Swell that has propagated into the forecast region is not considered in the parametric ap-

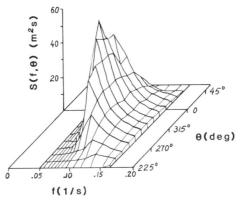

Figure 3.9. Computer model output showing a directional wave spectrum for Hurricane Eloise. (After Long, 1979.)

proach, but is added to the hybrid parametric model by solving the transport part of the balance equation:

$$\frac{\partial S}{\partial t} = - C_g \frac{\partial S}{\partial x}$$

over a grid. The hybrid parametric model thus combines the swell propagation aspects of the computerized spectral model and the ease of estimating spectral shape of the parametric approach. This type of model has been used effectively by Gunther et al. (1979), Weare and Worthington (1978), and Ewing et al. (1979) to predict severe waves in the North Sea for offshore design applications.

Although wave models have had some success in providing useful information for marine applications, important problems concerning the exact physics of the source and sink functions in the energy-balance equation are not well understood. Basic research still needs to be done on how winds cause wave growth, how component interaction can be included efficiently in wave models, and how waves decay. Also, it is important to remember that all wave models give outputs no better than the wind inputs. Thus, improvements in the measurement of winds at sea, possibly through remote sensing techniques, are needed to produce significant improvements in wave forecasting.

3.6. CALCULATION OF DESIGN WAVES

An important application of computer wave modeling is estimating extreme wave heights for site selection and design specifications of offshore structures. The high cost per unit height of large offshore structures must be

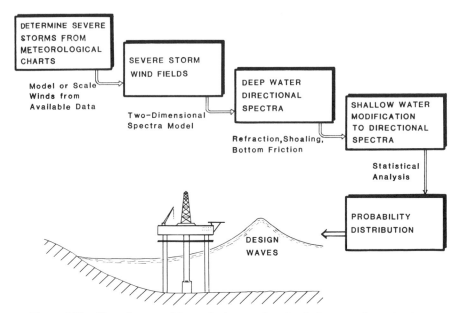

Figure 3.10.   Flow diagram of the analysis procedure for design-wave determination.

balanced against the loss that would occur from total failure. Design-wave conditions needed to make this cost–benefit analysis could be obtained from site-specific measurements covering a time period representative of the structure's designed life. In general, such a record of extreme wave observations will not be available. A statistical wave modeling analysis, as outlined in Figure 3.10, is therefore employed to calculate the required extreme wave climatology based on available wind data. Marine weather charts for the desired ocean area covering a period of 20 to 30 years are examined to select the most severe storms. Wind fields for each severe storm are developed for 6-hour intervals and are used as input to a computer wave prediction model such as the Wilson (1965), Cardone et al. (1976), or Weare and Worthington (1978) models. A directional spectral model is generally preferred for this type of calculation. For shallow water sites the model-generated deep-water-wave energy spectra must be modified for the shallow water effects of wave refraction, shoaling, and bottom friction. A numerical wave refraction computer program (see Dobson, 1967) which can be adjusted to include bottom friction effects (see Bretschneider and Reid, 1954) is used for such modifications. Detailed bottom topography information is needed for the wave refraction calculations. The refraction model produces a number of shallow water directional wave spectra, from which the required record of extreme wave heights can be computed directly. In very shallow water, wave heights are limited by breaking to approximately 0.78 of the water depth.

The final step in the procedure is a statistical analysis of the extreme wave heights calculated from the model. The largest waves calculated for the location of interest are tabulated and ranked from the lowest height value, designated as $i=1$, to the highest, designated as $i=N$. The letter N also represents the total number of extreme wave heights used in the analysis. The cumulative probability P_i of obtaining a wave height less than each of these ith ranked values is

$$P_i = \frac{i}{N+1}; \qquad i = 1, 2, 3, \ldots, N$$

For example, if the wave modeling analysis produced 10 extreme wave heights, then the probability of getting a wave height equal to or less than the first ($i = 1$) would be

$$P_i = \frac{1}{10+1} = \frac{1}{11}$$

and the probability of a wave height equal to or less than the last ($i = 10$) would be

$$P_i = \frac{10}{10+1} = \frac{10}{11}$$

Another parameter used in the in extreme-event analysis is the return period R. This is the average time interval between occurrences of given extreme wave heights. Return period can be related to the probability values of a statistical distribution by

$$R = \frac{1}{(1 - P_i)\left(\dfrac{N}{\text{Total Model Years}}\right)}$$

For example, if 30 years of data were used to develop 10 extreme wave heights at a given site the return period for the first, or lowest, height ($i=1$) would be

$$R = \frac{1}{\left(1 - \dfrac{1}{11}\right)\left(\dfrac{10}{30}\right)} \approx 3.3 \text{ years}$$

and the return period for the highest ($i=10$) would be

$$R = \frac{1}{\left(1 - \dfrac{10}{11}\right)\left(\dfrac{10}{30}\right)} \approx 33 \text{ years}$$

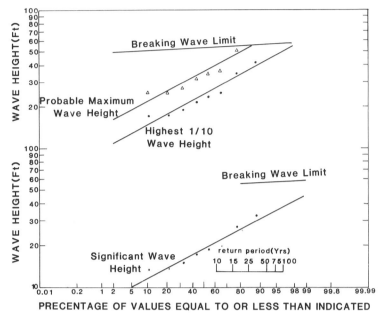

Figure 3.11. Wave height return period calculation for offshore light tower. (From Earle, 1975.)

The ranked wave heights and associated probabilities are generally plotted on probability paper. A "best fit" line through the extreme wave data can be extended to return periods longer than the period over which the wave model calculation has been made to obtain estimates of the 50- and 100-year extreme wave heights.

An example of a wave height probability and return period graph is shown in Figure 3.11. This plot represents a calculation of design waves for light towers located off the east coast of the United States. In the analysis of Earle (1975) 30 years of severe-storm data were used as input to a directional wave model. The probable maximum wave height, the highest 1/10 wave height, and the significant (highest 1/3) wave height were all calculated from the shallow water wave spectra. The heights are ranked by probability and return period and plotted on the probability graph. The 30 years of data used for this analysis are extended into longer return periods until limited by the wave breaking height (i.e., 0.78 of water depth). Similar modeling and statistical analysis procedures are used to predict extreme winds, currents, and storm surges for various offshore applications.

Four

TIDES, STORM SURGES, AND INTERNAL WAVES

In the previous chapter we gave a brief outline of ocean wind wave theory and predictive models. To complete this survey of ocean waves we now focus our attention on astronomical tides, storm surges, and internal waves. A knowledge of the astronomical tides, which produce the periodic rise and fall in coastal sea level, is important to safe coastal navigation. Storm surges, abnormal rises in coastal sea level caused by storms, produce coastal flooding, dictate design limitations of coastal engineering projects, and cause beach erosion. Internal waves, which form at subsurface density interfaces, affect the detection of objects by sonar.

4.1. THE ASTRONOMICAL TIDES

Astronomical tides are caused by the combined gravitational and centrifugal forces of the Earth, sun, and moon. Nonastronomical factors such as the shape of the coastline, bottom topography, and effects of storms may cause local modifications to the periodic rise and fall of the tides.

A useful model of astronomical tides can be developed by considering the solar equilibrium tide. In this model, as illustrated in Figure 4.1, the solar tide results from the relationship between gravitational and centrifugal forces. At the Earth's center the sun's attractive force G and the centrifugal force C due to the movement of the Earth in its orbit exactly balance ($G = C$). The sun's gravitational force decreases with distance from the sun, while the centrifugal force is constant over the Earth. On the side of the Earth farthest away from the sun the centrifugal force is larger than the gravita-

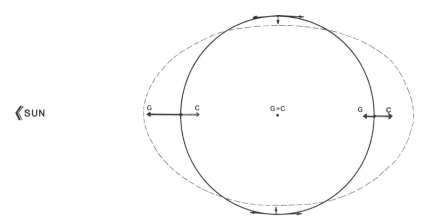

Figure 4.1. Tide-generating-force diagram, with G = gravitational attraction and C = centrifugal force, showing an exaggerated solar equilibrium tide as a dashed line.

tional force. This leads to a net force pointing away from the Earth's center (since $C > G$). On the side of the Earth closest to the sun gravitational force is greater than centrifugal force ($G > C$) and the net force will again point away from the Earth's center. The combined result of these differences in forces is a rise in sea level called the solar equilibrium tide. A similar difference of forces exists between the Earth and the moon, producing a lunar equilibrium tide. The tidal force of the moon is more than twice that of the sun because of the nearness of the Earth and moon compared with the distance between the Earth and the sun. Twice each month, at the times of the full and new moons, the tide-generating forces of the sun and moon reinforce each other. The resulting higher than normal tides are called spring tides. At the time of the quarter moon, when the Earth, sun, and moon are in quadrature, the solar and lunar tides are out of phase, and the smaller-amplitude neap tides result. Figure 4.2 shows the relative positions of the Earth, sun, and moon and the resulting tides at different times during the lunar month.

The astronomical tides, having wavelengths of about half the circumference of the globe, are by definition shallow water waves in the relatively shallow ocean basins. As the wave sweeps across the ocean, the resulting changes in sea elevation produce horizontal tidal currents which flow into a region as the crest of the tidal wave approaches and away from the region as the wave trough approaches. Flood tide occurs when currents are flowing into a region and sea level is rising, while ebb tide occurs when tidal currents are flowing out of a region and sea level is falling. The tidal stage between flood and ebb, when the tidal elevation is reversing, is called slack water.

The geometry of the ocean basins and the Coriolis force also affect the idealized equilibrium tidal wave. The open ocean tidal wave actually

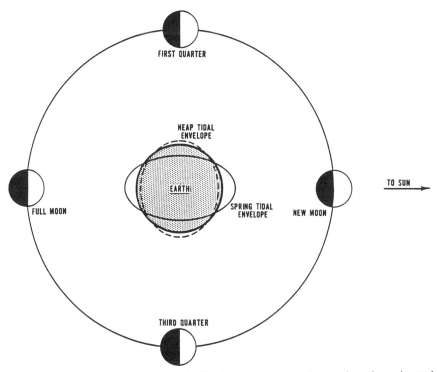

Figure 4.2. The relative positions of the Earth, sun, and moon that produce the spring and neap tides.

"sloshes" around the basin about a nodal, or amphidromic, point, much as tea would in a rotating cup. The result is a characteristic harmonic ocean tide with a typical range of only about half a meter. However, in some bays or tidal inlets resonance and convergence effects produce tidal ranges from 3 up to 15 m. Figure 4.3 shows typical elevation curves for several west coast ports of the United States. Note the larger than normal spring tidal elevations, especially at Anchorage, Alaska, during the time of new moon on the 20th day. Tidal elevations also change as the moon moves in its elliptical orbit around the Earth, with higher tides at perigee, when the moon is closest to the Earth, than at apogee, when the moon is farthest from the Earth. The declination of the moon above or below the plane of the Earth's equator also affects tidal amplitudes.

The series of tidal elevations at specific locations is produced by the periodic orbital motions of the moon and sun and modified by local factors. The method of harmonic analysis has been used to make tidal predictions from observations of tidal elevations. These predictions are based on the fact that a tidal record is composed of a finite number of tidal constituent

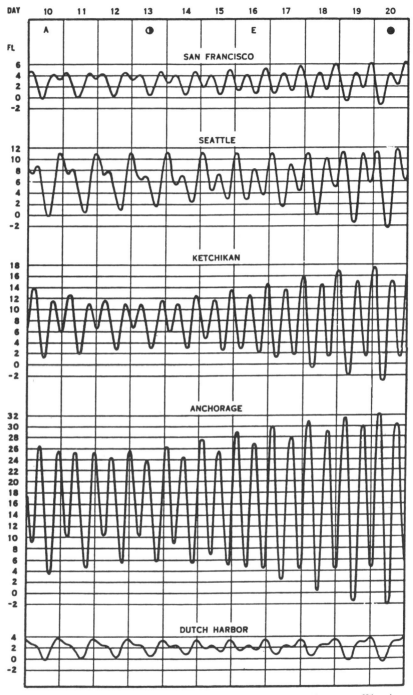

Figure 4.3. Typical tidal elevation curves for locations along the Pacific coast. (From U.S. Department of Commerce, 1970.)

68

components, each having its own period, phase angle, and amplitude. The period of each harmonic constituent is known from astronomy, while the local amplitudes and phase angles are obtained from the harmonic analysis of the tidal record. This leads to a harmonic representation of the local tide record as

$$\eta(t) = a_0 + \sum_{i=1}^{N} a_i \cos\left(\frac{2\pi}{T_i} t + \epsilon_i\right)$$

where $\eta(t)$ is the tidal elevation time series composed of N constituent harmonic components, each with an amplitude a_i, a phase angle ϵ_i, and a period, T_i. The mean value over the time series is a_0. Tidal prediction using the harmonic method is done in two steps. First, elevation data covering all important constituent periods are harmonically analyzed to determine the unknown phase angles and amplitudes of each constituent. Given this information and the known astronomical periods, the components are recombined mathematically to produce the tidal prediction. A detailed manual on using harmonic analysis for tidal predictions has been written by Schureman (1958), while a user's guide to a computer program for harmonic analysis of data at tidal frequencies can be found in Dennis and Long (1971).

Information on tidal elevations and tidal currents is important for many applications. Safe navigation of deep-draft supertankers requires accurate tidal elevation predictions to avoid shallow water reefs and prevent grounding. Tidal currents also affect these large ships as they navigate at slow speed in close quarters in harbors. Published tidal predictions for most of the major ports of the world are available to the mariner. In the United States the *Tide and Tidal Current Tables* are published annually by the National Oceanic and Atmospheric Administration. These tables give hourly predicted tidal heights and tidal currents at selected locations. Tidal information is also used to construct tidal datums which locate the seaward boundary of shorefront properties. This boundary is usually established at a specific water depth that is based on the mean low water tidal elevations observed over a long period of time. Military applications of tidal information include navigation of naval vessels, amphibious operations, and the setting of underwater pressure-activated explosive charges.

The harnessing of tidal energy has fascinated engineers for years. A dam built across the mouth of a suitable bay would make it feasible to control the tide and to use it to drive electrical turbines. In 1959 work was started on the world's first commercial tidal power plant. Built on the Rance Estuary near St. Malo, France, it has produced power since 1966. With the reduction in the availability of fossil fuels and the problems of nuclear power generation, the tides could be a valuable energy source, since tidal energy is renewable and involves proven engineering techniques. A more complete review of tidal power will be given in Chapter 10.

4.2.   COASTAL STORM SURGES

The term coastal storm surge is used to describe a departure from the astronomical tide produced by storms. Both tropical cyclones, formed near the ITCZ, and extratropical cyclones, formed near the polar front, can produce coastal storm surges. Severe surges have caused increases of coastal water level in excess of 8 m ($\sim$ 26 ft) on open coasts and even larger increases in bays and estuaries. Estimates of water level changes that occur during such storms are important for planning coastal engineering projects and for calculating nearshore flooding areas for evacuation planning.

Tropical cyclones are nearly circular storms with winds that may reach 75 m/s ($\sim$ 150 knots) revolving around a region of relative calm known as the central eye. The eye is the area of lowest atmospheric pressure. Pressure and wind speed increase rapidly outward from the eye to a zone of maximum wind speed some 10 to 100 km from the storm's center. This distance is called the radius to maximum wind. Wind speed generally decreases as one proceeds outward from this location. The central pressure is the best single index for estimating the storm surge from a tropical cyclone. Additional factors, such as the radius to maximum wind and the forward speed, are also used to characterize a storm's coastal surge potential. Severe tropical cyclones that affect the coastline of the United States are called hurricanes. Figure 4.4 illustrates the sequence of tidal elevation changes that occur during landfall of a hurricane.

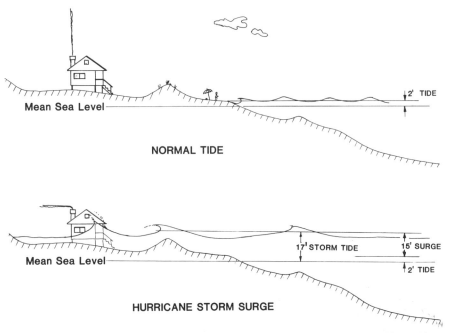

Figure 4.4.   Sequence of tidal elevations (in feet) during the passage of a hurricane. (After National Oceanic and Atmospheric Administration, 1979.)

Extratropical cyclones are generally larger than tropical cyclones, with maximum winds that may reach 25 m/s (~ 50 knots). These storms have produced storm surges over long regions of coastline. Extratropical storm surges may persist over more than one tidal cycle. This persistence, combined with the large coastal area of influence, may result in large-scale coastal erosion. In the United States extratropical cyclones have produced destructive storm surges mainly during the winter months from November to April, whereas the hurricane season is from June to November.

A cyclone produces a storm surge through a complicated interaction of meterological, oceanographic, and hydrodynamical processes. These processes have been divided into five major components.

Pressure setup.   Cyclones, especially tropical cyclones, cause a rise in sea level due to an inverted barometer effect which occurs as the water level rises slightly under low-atmospheric-pressure regions. In theory one can compute this change in water elevation $\Delta\zeta$ using a form of the hydrostatic equation if given the pressure change Δp from the outside to the inside of the storm:

$$\Delta\zeta = \frac{|\Delta p|}{\rho g}$$

where ρ is the water density and g is gravity. This leads to the approximation that for each millibar of pressure change the sea level changes 1 cm:

$$\Delta\zeta = \frac{10^3 \text{ dyne/cm}^2}{(1.02 \text{ g/cm}^3)(980 \text{ cm/s}^2)} \approx 1 \text{ cm}$$

The pressure setup effect is larger and more concentrated in well-developed tropical cyclones than in extratropical cyclones. In Hurricane Camille, a 1969 storm that went ashore in the Gulf of Mexico, Δp was 108 millibars, which produced a theoretical pressure setup of approximately 1 m (~ 3 ft).

Direct transport.   The wind component perpendicular to the coast produces a direct wind-driven transport which increases the coastal sea level. This effect is proportional to the magnitude of the surface wind stress.

Ekman transport.   The Earth's rotation causes a deflection in wind-driven currents to the right in the Northern Hemisphere and to the left in the Southern Hemisphere. As a result of this deflection, the wind stress component parallel to the coast, with the coast to the right of the wind direction in the Northern Hemisphere and to the left in the Southern Hemisphere, will increase coastal sea level. This effect is proportional to the magnitude of the surface wind stress.

Wave setup.   Wind waves moving into shallow water produce an increase in nearshore water elevation, especially as the waves break on the beach. This "wave setup" process has a maximum effect at open coastlines and in regions where the depth increases rapidly with distance from the

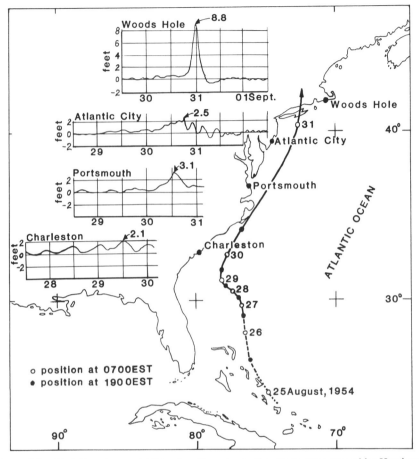

Figure 4.5.   Variations in coastal tide elevations due to the storm surge caused by Hurricane Carol in 1954. (After Harris, 1963.)

shore. This allows large waves to approach close to the beach before break-ing. Under favorable conditions this process may account for an increase in coastal sea level elevation of as much as 2 m (~ 6 ft).

Precipitation.   Tropical cyclones have produced as much as 30 cm (~ 1 ft) of rain in 24 hours, resulting in flooded rivers and streams that increase the water elevations near the heads of bays and estuaries.

Storm surges may also be augmented by the astronomical tide. Thus, this effect must be included in coastal storm surge predictions. Figure 4.5 shows a typical sequence of storm-surge-generated tidal elevations that occurred following the passage of a tropical cyclone along the east coast of the United States. Note that the tidal elevation was highest at the Woods Hole observa-tion site, where pressure setup, direct transport, and wave setup combined to produce higher elevations than were observed at the Charleston, Portsmouth or Atlantic City observation sites.

4.2.1   The Nomograph Method for Estimating Coastal Storm Surge

A method to obtain a first approximation for the peak storm surge from tropical cyclones has been developed by Jelesnianski (1972). He combined empirical data and a numerical storm surge model to produce a set of nomographs that permit rapid estimates of surge height to be made for coastal regions of the United States. The approach can be applied to locations that may require evacuation from coastal storm surges.

Figure 4.6(a) is a nomograph that is used to compute the peak coastal storm surge during a hurricane. The peak open ocean surge, S_I, is given as a function of the variables Δp (the central pressure deficit) and R (the radius to maximum wind). The nomograph was constructed for a hurricane moving perpendicular to the coastline at a speed of 15 mph (~ 6.7 m/s). Under these conditions the highest surge elevations occur with larger values of Δp for the $R = 30$ statute miles (~ 48 km). The value of surge elevation is also adjusted for the effects of the local bottom topography with a shoaling factor F_s. Figure 4.6(b) shows the computed shoaling factor for coastal locations in the Gulf of Mexico. The maximum surge is further modified and corrected for the storm's speed V_F and direction of motion ψ relative to the coastline. This is done by using a correction factor for storm motion F_M, as shown in Figure 4.6(c). The corrected peak storm surge S_P is computed as

$$S_p = S_I \cdot F_s \cdot F_M$$

To illustrate the use of the nomograph method, we will calculate the peak surge produced by Hurricane Camille. Hurricane Camille went ashore with $\Delta p = 108$ millibars, and $R = 15.6$ statute miles (~ 25 km). From Figure 4.6(a) the peak unadjusted surge $S_I = 22$ ft (~ 6.7 m) above mean sea level and from Figure 4.6(b), based on the landfall location (west of Biloxi, Mississippi), $F_s = 1.23$. The storm approached the coastline at an angle $\psi = 102°$ with a speed $V_F \simeq 14.5$ mph (~ 6.5 m/s), so $F_M = 0.96$. The peak surge using the Jelesnianski nomograph method is thus

$$S_p = (22 \text{ ft}) \cdot (1.23) \cdot (0.96) = 26.2 \text{ ft } (\sim 8 \text{ m})$$

The observed peak surge of 24.2 ft (~ 7.4 m) above mean sea level occurred at Pass Christian, Mississippi.

4.2.2.   Empirical Forecasts of Beach Erosion

Beach erosion, characterized by a retreating beach, has had serious economic consequences for many coastal communities. Empirical techniques

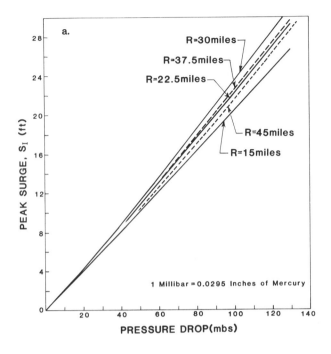

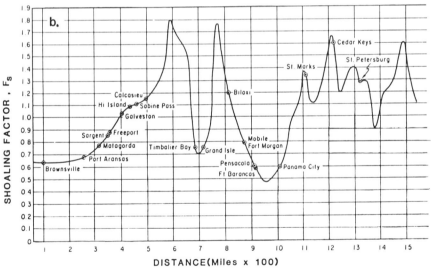

Figure 4.6. Peak storm surge (S_I) (*a*), shoaling factor (F_s) (*b*), and storm motion correction factor (F_M) (*c*) nomographs used with the storm surge calculation method developed by Jelesnianski. Peak surge is in feet, pressure drop in millibars (mbs), and distance in statute miles. (After U.S. Army *Shore Protection Manual*, 1973.)

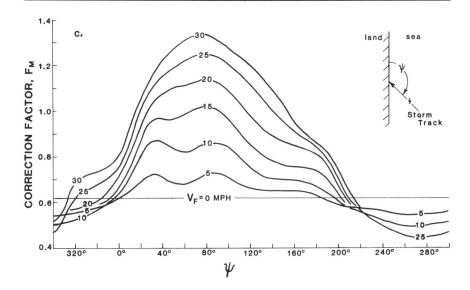

have proven useful for qualitative estimation of beach erosion caused by extratropical cyclones. One such technique was developed by the National Weather Service for the east coast of the United States and was used by Richardson (1978) to make operational beach erosion forecasts. This technique is based on the beach erosion intensity scale shown in Figure 4.7. The erosion intensity relationship was developed by taking reported accounts of beach erosion severity and then fitting appropriate meteorological and oceanographic parameters to the scale. The analysis indicated that the four major factors in the erosion process were the maximum observed tide (astronomical plus storm surge), the maximum storm surge height alone, the generalized storm duration (number of high tides for which the critical value of 2.5 ft ($\sim$ 0.8 m) for surge and astronomical tide is reached or exceeded over a long coastal range), and the variable storm duration (computed like the generalized storm duration, but with the critical value changing from location to location along the coast).

The observed erosion intensity was correlated with these four predictors, leading to the empirical beach erosion forecast equation

$$BE = -0.23 + 1.44t_g + 0.13S_p^2 + 0.70t_v + 0.23T_h$$

where BE is the beach erosion scale (from 0 to 16), t_g is the generalized storm duration, S_p is the maximum storm surge (in feet), t_v is the variable storm duration, and T_h is the maximum tidal height above mean sea level (in feet). This empirical equation was verified using data from six extratropical storms

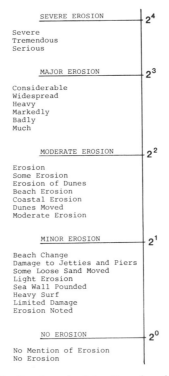

Figure 4.7. Beach erosion intensity scale and associated qualitative descriptive terms. (After Richardson, 1978.)

that caused major to severe erosion during the 1977–1978 winter season, and has been used operationally to make beach erosion forecasts.

4.3 INTERNAL WAVES IN THE OCEAN

Internal waves are subsurface undulations that propagate over ocean layers of unequal density. These waves cover a spectrum of wave periods from tidal periods to periods of minutes. In the open ocean internal waves are frequently found along the main thermocline moving as progressive waves with wavelengths of up to several kilometers. The velocity of internal waves is usually expressed as

$$C = \sqrt{gh\left(\frac{\rho - \rho'}{\rho}\right)}$$

where h is the depth of the density interface (usually the thermocline), g is gravity, ρ' is the density of the upper fluid, and ρ is the density of the lower

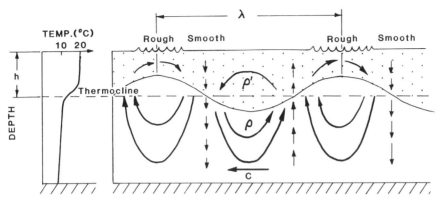

Figure 4.8.   A progressive internal wave moving along a thermocline of depth h from right to left with a speed c and wavelength λ.

fluid. The primary generating mechanisms for internal waves are tidal action, flow over irregular bottom topography, instabilities near water mass boundaries, and tropical cyclones.

Figure 4.8 illustrates an idealized internal wave and the resulting surface pattern. These patterns have been observed regularly by ships and aircraft at sea. The rough portion of the pattern is thought to be caused by an increase in the speed of the water just below the surface as a wave ridge replaces a wave trough. This causes water to be forced to pass through the constricted region between the ocean surface and the raised thermocline. The smooth area, or surface slick, occurs in the convergence zone above the downward flow between adjacent cells.

Although internal waves cannot be observed directly from satellites, their surface manifestations have been observed as alternating rough and slick areas in the ocean. In calm ocean waters where sun glint, which is the area of the sun's reflection that can be observed by satellite, is present, internal waves have been observed as distinct alternate bright and dark bands on the sea surface. Figure 4.9 is a satellite image of the Sulu Sea, an ocean region in the western Pacific Ocean between the southwest Phillipines and Borneo. The surface manifestation of an internal wave pattern can easily be seen. The image suggests a pattern of internal waves radiating from a passage in the Sulu Archipelago with at least five separate packets. The longest wavelength in packet 3 is ~ 7 km, with the wavelength decreasing from the front to the rear of the packet.

Internal waves moving along the thermocline have an important influence on underwater sound transmission (Katz, 1967). As the thermocline oscillates, sound waves intersect it at varying angles, resulting in a disruption in acoustical propagation. Because of this disruption sonar may be affected adversely in areas of internal waves. Acoustical effects may be such that false targets are detected while actual targets pass undetected. Internal

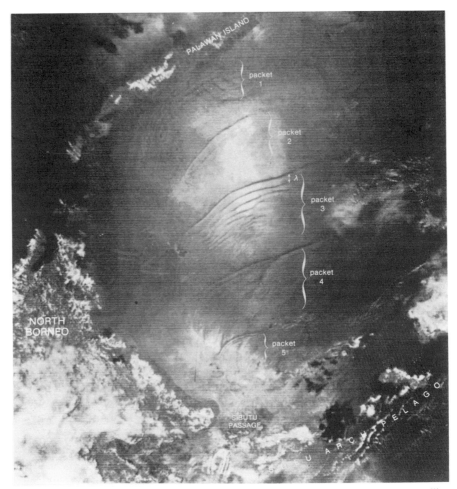

Figure 4.9.   Enlarged view of internal wave packets in the Sulu Sea as observed on satellite imagery. (From U.S. Navy *Tactical Applications Guide*, 1981.)

waves have also been suspected of being a danger to submarine operations. For example, a submarine passing through the upward current of an internal wave would have to change ballast to maintain a constant depth and would then find itself diving in the downward current portion of the wave. Under these circumstances, it is possible that the submarine might exceed its structural limitations. This situation has been considered as a possible cause for the loss of the *USS Thresher* (Larson et al., 1971).

Five

CONTAMINANT ADVECTION
AND MIXING

One area of oceanography that has been applied to a number of contemporary problems is the modeling of contaminant advection and turbulent diffusion. Problems ranging from the fate of oil spilled at sea to the long-term ecological effects of ocean dumping require an understanding of the mechanisms by which pollutants are transported and mixed in the ocean. Chapter 2 presented a brief outline of the equations representing the turbulent exchange of momentum in the ocean. A similar process, turbulent diffusion, is responsible for the mixing of contaminants. In this chapter an environmental assessment of a coastal outfall project illustrates the application of the modeling of advection and turbulent diffusion processes to a coastal pollution problem.

5.1. ADVECTION AND TURBULENT MIXING

When a contaminant such as an oil spill or a chemical discharge is introduced into the marine environment, it will be advected away from its source by the average large-scale current while at the same time undergoing mixing by small-scale turbulent eddies. The mixing process causes the initial contaminant patch to grow with time, thereby reducing concentrations within the patch. As the patch continues to grow, the size of the eddies that can act on it also increases. In general, eddies larger than the patch act to advect the contaminant away from its source, while eddies the same size as or smaller than it produce turbulent mixing. Turbulent mixing, or diffusion, increases with time in proportion to patch size. Mathematical models based on actual measurements relate this increase in patch size to the consequent decrease in peak contaminant concentration.

The governing equation used in calculations of the advection and turbulent diffusion of contaminants considers the average concentration of the contaminant $\bar{c}$ and the fluctuations of contaminant concentration c' about the average value caused by turbulent eddies. The governing equation takes the form

$$\frac{\partial \bar{c}}{\partial t} = - \left[\bar{u} \frac{\partial \bar{c}}{\partial x} + \bar{v} \frac{\partial \bar{c}}{\partial y} + \bar{w} \frac{\partial \bar{c}}{\partial z} + \frac{\partial}{\partial x}(\overline{c'u'}) + \frac{\partial}{\partial y}(\overline{c'v'}) + \frac{\partial}{\partial z}(\overline{c'w'}) \right]$$

where the overbar indicates average values and the prime indicates turbulent fluctuations. This equation may be expressed in terms of excess concentration above some ambient or background level such that $\Delta c = \bar{c} - c_a$, where c_a is the ambient concentration. The turbulent diffusion terms on the right hand side of the equation are generally expressed using constant eddy-diffusion coefficients, which are analogous to the eddy viscosity coefficients discussed in connection with the turbulent transfer of momentum. With these modifications the governing equation of oceanic advection and diffusion is rewritten as

$$\frac{\partial \Delta c}{\partial t} = - \left[\bar{u} \frac{\partial \Delta c}{\partial x} + \bar{v} \frac{\partial \Delta c}{\partial y} + \bar{w} \frac{\partial \Delta c}{\partial z} \right] + E_x \frac{\partial^2 \Delta c}{\partial x^2} + E_y \frac{\partial^2 \Delta c}{\partial y^2} + E_z \frac{\partial^2 \Delta c}{\partial z^2}$$

where E_x, E_y, and E_z are constant turbulent-eddy diffusion coefficients having dimensions L^2T. In this equation the term $\partial \Delta c / \partial t$ represents the local change in concentration at one point in space. The first three terms on the right-hand side of the equation represent the advection of the contaminant by the average current components ($\bar{u}$, $\bar{v}$, $\bar{w}$) and the last three terms on the right-hand side represent the horizontal and vertical diffusion of the contaminant by turbulent eddies.

A contaminant may be introduced into the marine environment as a suspension of particles or as a solution of different density from that of the receiving waters. If in particulate form, the contaminant cloud will undergo gravitational settling in addition to advection and turbulent diffusion. If the contaminant is less dense than the receiving waters it will rise due to bouyancy. These rates can be easily estimated given the difference in water and particulate density. It is also possible for a contaminant to undergo chemical decay with time. This decay can be included in the governing equation by assuming that the rate of decay is proportional to the concentration of material present at a given time:

$$\frac{\partial \Delta c}{\partial t} = - K \Delta c$$

where the decay constant K has dimensions 1/T. This relationship leads to an exponential decay in the local concentration with time.

In summary, a contaminant dumped into the ocean will undergo advection by the mean current and diffusion by turbulent eddies superimposed on the mean flow, it will rise or sink depending on its density relative to the density of ambient ocean waters, and it may undergo decay.

5.2. BRINE DISPOSAL IN THE GULF OF MEXICO

In a number of pollution assessment applications it is necessary to estimate the extent of a plume discharged from a given source. Assessments of by-products from oil drilling operations, thermal plumes from nearshore power plants, ocean mining discharges, and discharges from coastal sewage out-falls all require such calculations. Simple and accurate models have been developed for predicting the temporal and spatial distribution of such contaminant plumes.

Such a model was used by the United States government in the Strategic Petroleum Reserve (SPR) program. The oil reserve project was initiated after the oil embargo of 1973 to improve the nation's strategic position by the storing of up to 750 million barrels of oil in subterranean salt domes. These geological formations are located along the Gulf of Mexico coast near oil distribution and refinery facilities. Large quantities of fresh water were used to dissolve and rinse out the salt deposits, eventually producing large sub-surface regions to store the oil. A by-product of this operation was a saturated brine solution with a salinity of 260 to 280 parts per thousand (ppt), compared with the average Gulf of Mexico salinity of $\sim$ 35 ppt. The brine solution is removed from the cavern as it is displaced by the incoming oil. The initial plan called for disposal of about 500,000 to 1,000,000 barrels of brine per day over 18 months at each of seven sites.

The government initiated an assessment to determine the environmental impact of disposing of the brine into the biologically productive waters of the Gulf of Mexico. Plans called for the brine to be pumped from the salt domes and along the ocean bottom a number of miles offshore. The brine would then be jetted upward into the water by a specially designed diffuser system which would actually be an approximately 1000-ft ($\sim$ 300 m) continuation of the offshore pipeline with about 50 vertical nozzles for spraying the brine into the receiving waters.

5.2.1. Modeling of Brine Disposal

The assessment of the environmental consequences of the proposed brine disposal operation was based on the transient plume model described in detail by Adams et al. (1975). This model divides the total impact zone into three analysis regions—the near field, the intermediate field, and the far field—based on the dominant physical processes in each region. In the near-

Near Field

Dilution ⟶ Diffuser Design & Current

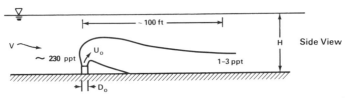

Figure 5.1.   Near-field region of the transient plume model. (After National Oceanic and Atmospheric Administration, 1977a.)

field region, illustrated in Figure 5.1, dilution of the brine effluent is governed by the spraying action of the diffuser nozzles and by the current speed near the sea floor. The diffuser design used to produce a desired near-field concentration level depends on the discharge flow rate, Q_0, the discharge excess salinity concentration Δc, and the discharge density ρ_0. The discharge density can be combined with the water density to define a discharge relative density difference:

$$\frac{\Delta\rho}{\rho} = \frac{\rho_0 - \rho}{\rho}$$

In the SPR model application, worst-case conditions of flow rate $Q_0 = 1,100,000$ barrels of brine per day, excess salinity $\Delta c = 230$ ppt, and relative density difference $\Delta\rho/\rho = 0.25$ were used.

Diffuser design parameters such as nozzle discharge speed U_0 and nozzle diameter D_0 were combined to form a nondimensional discharge Froude number given as

$$F_0 = \frac{U_0}{\sqrt{\dfrac{\Delta\rho}{\rho}\, g\, D_0}}$$

where g is gravity. This nondimensional parameter is of specific interest, since it is used to compare available experimental data on diffuser dilution efficiency and diffuser design specifications. The desired dilution level requires a Froude number of about 18, which corresponds to diffuser specifications that are easily met. For this F_0 value the near-field dilution D_l is related to the bottom current speed V by the linear equation

$$D_l = 45 + 55V$$

Intermediate Field

Dilution $\longrightarrow$ Lateral Spreading & Diffusion

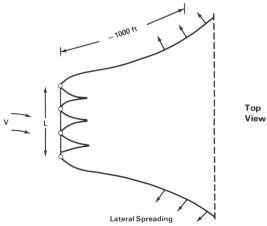

Figure 5.2. Intermediate-field region of the transient plume model. (After National Oceanic and Atmospheric Administration, 1977a.)

where V is in ft/s. For a typical bottom current of 1.0 ft/s ($\sim$ 30 cm/s) a near-field dilution by a factor of about 100 could be expected with a diffuser designed for $F_0 = 18$.

At the end of the near-field region, or about 100 ft ($\sim$ 30 m) downcurrent from the diffuser, is the start of the intermediate field. In this region, illustrated in Figure 5.2, discharges from individual diffuser nozzles merge to form a single plume. The transient plume model is structured to allow intermediate-field dilution to occur until the lateral spreading of the plume is balanced by vertical diffusion. Intermediate-field dilution is related to the overall length of the diffuser. Adequate length allows the plume to spread over a large enough area that a wedge of high-density brine would be inhibited from forming on the sea floor. Experimental data indicate that this requires the diffuser length L to satisfy the inequality

$$L > \frac{Q_0 \, \Delta \rho g}{\rho V^3}$$

which means a diffuser length of 3000 to 6000 ft ($\sim$ 900 to 1800 m) is needed for this project, even for abnormally low current speeds.

The intermediate field ends and the far field begins about 1000 feet ($\sim$ 300 m) downcurrent from the diffuser. In the far field, illustrated in Figure 5.3, the final dilution phase is governed by the ambient processes of advection and diffusion. These processes are modeled by superposition of a number of

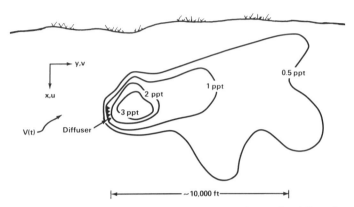

Far Field

Dilution ⟶ Advection & Diffusion

Figure 5.3. Far-field region of the transient plume model. (After National Oceanic and Atmospheric Administration, 1977a.)

contaminant patches that leave the intermediate field and enter the far-field region. Dilution in the far field is calculated from the equations given in Section 5.1. The far-field region extends ~ 10,000 feet (~ 3 km) or more downcurrent from the disposal site.

In the transient plume model the contaminant is tracked using a computer solution on a numerical grid. The governing equations in each of the three regions are solved at each grid point for specific time intervals. The environmental inputs to the model are a time series of bottom currents and the eddy diffusion coefficients. The horizontal mixing coefficients (E_x and E_y) were estimated from the work of Okubo (1962), which presented relationships between the increase in the contaminant patch with time and these coefficients. The vertical mixing coefficient (E_z) is usually assumed constant in the transient plume model.

Since bottom current data, which serve as key input to the model, were not available for any of the proposed brine disposal sites, a synthetic time series of bottom currents was developed to provide the necessary information. A statistical process called a Markov model, in which current speed at one time is correlated with the speed at previous times, was used. Inputs to the Markov simulation could be developed from basic meterological and oceanographic data that were readily available for each site. These data include the amplitude and period of the local tidal current, the mean number of storms passing near the site, storm intensity, the approximate amplitude of the resulting wind-driven currents, and the velocity of the permanent coastal current. Figure 5.4 shows a typical Markov current model of the

Use of Synthetic Current

$$V = V_T + V_{NT} + V_{WD}$$

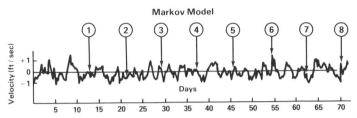

Figure 5.4. Synthetic current time series showing sampling points. Components of the synthetic current are the tidal current V_T, the permanent current V_{NT}, and the wind-driven current V_{WD}. (After National Oceanic and Atmospheric Administration, 1977a.)

local bottom current which was used to generate outputs of the transient plume model. Actual measurements later proved that this current simulation procedure was accurate enough for its intended purpose.

5.2.2. Application of the Transient Plume Model

The transient plume model provided key information for making a pre-disposal assessment of the environmental consequences of the brine disposal project. The model was used to plan the exact locations of the disposal sites, as an aid to diffuser design, and to estimate the space and time scales of the brine plume. This information was used by project scientists to evaluate possible ecological effects of the disposal operation on the valuable fishing grounds in the Gulf of Mexico. In turn, government officials were able to assess if the projected environmental costs were acceptable before the project was allowed to proceed.

Model results that were of particular interest in the environmental assessment were "snapshots" showing areas covered by specific concentration levels at given times, plots of excess concentration versus area, and plots of excess concentration versus time. Since the brine plume was denser than the receiving waters, brine concentration decreased rapidly from the sea floor upward. Therefore the plots used in the assessment were mainly based on model outputs near the sea floor. The snapshots were especially useful in outlining the geographic area along the sea floor that could be covered by the brine plume. Figure 5.5 shows a typical plume pattern for a snapshot taken at time (8) on the Markov current simulation given in Figure 5.4. Note that

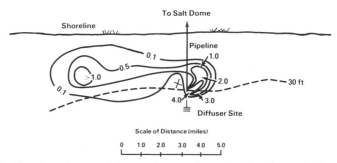

Figure 5.5. Transient plume model output for sampling point (8) on the synthetic current time series shown in Figure 5.4. Contour numbers are excess salinity in parts per thousand (ppt) and distance is in statute miles. (After National Oceanic and Atmospheric Administration, 1977a.)

the geographic scale is on the order of a few miles and that the dominant longshore current direction is reflected in the orientation of the plume.

Figure 5.6 illustrates an excess concentration versus area chart, also for the current sequence shown in Figure 5.4. The excess salinity contour $\Delta c = 1$ ppt covered an area along the sea floor of approximately 3×10^7 to 10^8 ft^2 (2.8×10^6 to 9.3×10^6 m^2) and the excess salinity contour $\Delta c = 3$ ppt covered an area of approximately 10^6 to 10^7 ft^2 (9.3×10^4 to 9.3×10^5 m^2).

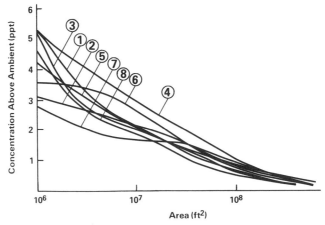

Figure 5.6. Excess salinity concentration versus area chart derived from the output of the transient plume model at the eight sampling points shown in Figure 5.4. (After National Oceanic and Atmospheric Administration, 1977a.)

Concentration vs Time Charts

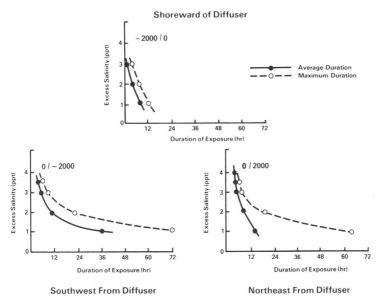

Figure 5.7. Excess salinity concentration versus time charts derived from the output of the transient plume model using the current time series shown in Figure 5.4. (After National Oceanic and Atmospheric Administration, 1977a.)

This model output was particularly useful in relating brine concentrations to the extent of possible ecological effects on stationary organisms that inhabit the sea floor.

Figure 5.7 indicates how modeled brine concentrations typically changed with time at locations 2000 ft (~ 600 m) shoreward, 2000 ft downcurrent (southwest), and 2000 ft upcurrent (northeast) of the diffuser. The effect of a southwestward coastal current is reflected in the model outputs, since the excess salinity concentrations are higher over longer durations in that direction than in the upcurrent direction. This output was particularly useful in relating the projected brine concentration to the effects on floating organisms. The output indicated that, in general, these organisms would receive only limited exposure to high concentration levels, except during current reversals.

These three types of model outputs allowed project scientists to define the spatial and temporal scale of the brine plume. The next step in the assessment procedure was to develop relationships between model outputs and local ecological effects. Although no well-documented limits of tolerance to hypersaline conditions are known for a wide variety of specific organisms,

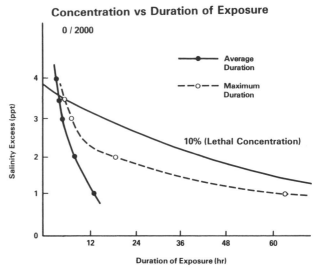

Figure 5.8. Overlay chart used to scale the ecological effect of the modeled brine plume at a point 2000 ft (~600 m) from the diffuser. The region above the solid line represents the concentration level and duration combination that was lethal to 10 percent of the test organisms. (After National Oceanic and Atmospheric Administration, 1977a.)

certainly each organism has an optimum range of salinity for maximum growth and reproduction. Additionally, the abruptness and duration of salinity change were assumed to be of particular importance in assessing ecological effects. Based on this logic and on modeled excess salinity levels, the main ecological effects were expected to be in the near-field region and only on relatively sensitive organisms. To get a more specific indication of the ecological effect, model outputs were combined with laboratory data on the effects of excess salinity concentrations on selected organisms. Figure 5.8 illustrates the approach employed. Model outputs were plotted on the same chart as the laboratory data. In the diagram data for two of the most sensitive organisms, shrimp larvae and spotted sea trout eggs, are plotted. This overlay approach led to the conclusion that at a distance of over 2000 ft (~ 600 m) from the diffuser fewer than 10 percent of the most sensitive local organisms would receive a lethal dose of brine.

After reviewing several combinations of model outputs and ecological effect data the Environmental Protection Agency (EPA) granted a permit to initiate brine disposal operations at a test site off the Texas coast. As a check to the model results the EPA required observations of salinity to be taken during disposal operations. Figure 5.9 illustrates a measured brine plume along with the observed current at the disposal site. Observations of brine plumes during the initial months of disposal operations were compared with model outputs made with the observed current as input. This comparison

March 22, 1980
Ambient Salinity (33.2)

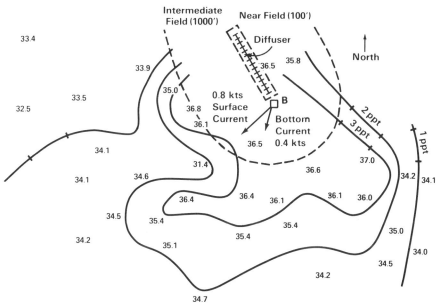

Figure 5.9. Observed salinity distribution near the diffuser during actual brine disposal operations. Numbers on the diagram are observed salinity and excess salinity in parts per thousand (ppt). (After National Oceanic and Atmospheric Administration, 1977a.)

indicated that the plume model and the Markov current simulation were effective in determining actual conditions.

The SPR environmental assessment project illustrates the application of oceanography to the decisionmaking process. A synthetic current time series and estimated diffusion coefficients provided the needed input for the transient plume model. Model outputs gave a crude but reasonable pre-disposal estimate of the dimensions of the brine plume that was needed for projecting ecological effects. This information was used by government officials to make project adjustments. Model outputs compared well with actual observations during disposal operations, and were below the level at which unacceptable damage to the marine environment would likely occur.

TOPICS IN APPLIED OCEANOGRAPHY

As the global demand for goods and services increases with world population, the oceans are being used more extensively for food, energy, minerals, transportation and pollutant disposal. Decisions concerning the location, timing, level of operations, and pollution control measures for each ocean use require an analysis of the benefits of each activity weighed against its environmental costs. These environmental assessments, designed to resolve potential ocean use conflicts, require adequate information concerning:

1. The spatial and temporal distributions of existing and proposed ocean use activities.
2. The spatial and temporal distributions of the pollutants that each ocean use activity will generate.
3. An understanding of the physical, ecological, economic, and governmental policy factors that collectively form the total environmental system under analysis.

Applied oceanography is defined as the development of the analysis techniques needed for the systematic assessment of a particular ocean use activity. It therefore represents the blending of available scientific knowledge into the needs of the operational ocean-user community. The physical environment, or the study of physical oceanography, constitutes but one component of the total system that must be considered. Applied oceanography also includes important ecological, economic, and governmental policy components. The three categories of applied oceanography to be addressed in the following chapters are:

Marine pollution, which represents the environmental costs of increased ocean use.

Marine resources, which represents the benefits society can obtain from the ocean.

Marine transportation, which is the use of the oceans for transportation and the effects of the oceans on maritime activities.

MARINE POLLUTION

Marine pollution is the introduction of impurities into the ocean, and includes oil spills from ships and exploration rigs, discharges from marine
mining and ocean engineering operations, and the dumping of domestic,
industrial, agricultural, and chemical waste. The consequences of marine
pollution range from short-term economic losses due to the unsightly fouling
of beaches by oil spills, garbage, and other floatable objects to less visible
but longer-term effects on the total ecological system.

The oceans cover more than 70 percent of the global surface, to a mean
depth of over 3.5 km. One question that often arises is: How can pollution
significantly modify such a large volume of water? The answer lies in the fact
that most of the accidental and deliberate incidents of marine pollution occur
in the shallow waters of the world's coastal oceans. This has led to obvious
ocean use conflicts because of the co-location of important ecological and
economic resources with pollutant sources. In response to this problem
some governments have initiated policies to monitor pollution, especially
along and near their coasts.

To organize our discussion of marine pollution we propose a total pollution management system that has four components. These are pollutant fate,
ecological effect, and economic impact and the associated governmental
policies. Pollutant fate is determined by meterological and oceanographic
processes that tend to modify the pollutant's concentration over time and
space. The ecological component is defined as the effects that pollutant
concentration levels have on marine organisms. In similar fashion, the economic impact of pollution is related to concentration levels. For example, a
major oil spill may cause direct economic losses by covering a recreational
beach or a more indirect, but just as damaging, impact due to its long-term

ecological effect on local commercial fish catches. The fourth component in this total system model, govermental policy, develops in response to the public's concern about the concentration levels, ecological effects, and economic impacts of marine pollution. This concern is often reflected in the policy actions of government officials.

The use of the fate, effect, impact, and policy system to address the complex problem of marine pollution has a number of advantages over the more traditional component-by-component treatment of the problem. These advantages include:

1. The ability to set research goals based on the needs of the total system.
2. The development of component models that have consistent spatial and temporal representations such that important systemwide questions can be addressed and necessary links between system components be developed.
3. Allowing critical terms, such as monitoring and assessment, to be defined with respect to the total system rather than in terms of only one or two of the system's components.
4. Showing physical oceanographers, ecologists, economists, and government policymakers how their work fits into a total marine pollution management system.

In this section a holistic approach to marine pollution is presented. Specific topics of discussion include oil spills at sea, ocean dumping, pollution monitoring, and marine pollution assessment. Additional information on the holistic approach to marine pollution can be found in Williams (1979).

Six

OIL SPILLS AT SEA

In an attempt to meet the increasing need for energy the search for petroleum at sea is becoming the largest oil exploration endeavor in the history of the industry. Initial interest in offshore oil followed the discovery of large oil fields just seaward of existing shoreline deposits. The technology was then developed to allow expansion into these fields, and soon oil was being extracted from under deeper and deeper waters. Pipelines and tankers carry this new oil from the well to onshore refinery and distribution facilities. This increased movement of oil at sea has not occurred without environmental costs. Oil wells themselves are a source of long-term chronic pollution from continuous small spills, while a number of massive oil spills have resulted from oil well blowouts and from collisions, groundings, and structural failures of oil tankers.

Ecologists have a general understanding of the ways in which oil affects marine organisms, including direct poisoning, smothering, genetic damage, and modification of the food supply. Such general effects can be correlated with the type and volume of oil spilled and to the duration of contact. However, comprehensive information on the effects of oil on the total marine ecosystem is, for the most part, unavailable. Data are either site or organism specific, and have generally been obtained from controlled laboratory experiments which may not be representative of actual conditions at sea. The economic impact of oil spills takes the form of soiled boats, fouled fishing gear, beaches, and seawalls, and contaminated fish stocks, marine mammals, and seabirds. The market value of impacts such as the cost of the lost oil, cleanup expenses, and revenue losses to commercially valuable beaches and local fish stocks can be estimated. The costs for nonmarket items such as seabirds, marine mammals, and damaged plant life are more difficult, if not impossible, to determine.

The emphasis in this chapter will be on the physical fate of oil spills. This information leads to an understanding of the concentration distribution that results from a given oil spill at sea. It is important to put such knowledge in proper perspective, since without a sufficient understanding of the ecological effect, the economic impact, and the implications for government policy that arise from specific concentration levels, a holistic assessment of the oil pollution problem will not be possible.

6.1. THE FATE OF OIL AT SEA

Environmental processes that determine the fate of oil at sea can be divided into two classes. The first includes processes that occur without regard to the size of the oil spill, such as the breaking up of the spill into patches by the action of wind and waves. These patches are then advected by the current and spread apart by turbulent diffusion. The second includes processes that are governed by the type and volume of oil spilled as well as by environmental conditions. Specific processes that influence the fate of oil at sea, as illustrated in Figure 6.1, are as follows.

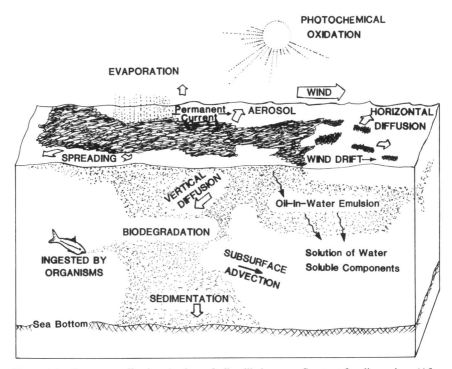

Figure 6.1. Processes affecting the fate of oil spilled at sea. See text for discussion. (After Kopenski and Long, 1981.)

Advection.   Once oil is spilled at sea it will be transported, or advected, away from the spill site by currents. The two commonly used components in oil spill advection calculations are the seasonal permanent current and the transient wind-drift current. Atlas charts or large-scale ocean current models are employed to derive the permanent current field. Transient wind-drift currents are computed from available wind records, usually by the wind factor method. Tidal currents are not used in advection calculations except for regions very near "target" areas because they are periodic and will lead to little or no net advection over a tidal cycle.

Turbulent diffusion.   Winds and wind waves cause large oil spills to break up into a number of small, distinct patches of oil which are then advected away from the spill site by the average current and diffused by the turbulent eddies in the water. Horizontal eddy diffusion causes oil patches to move slowly apart with time. Vertical diffusion is far less effective in mixing oil into subsurface waters.

Subsurface advection.   Oil spilled at the sea surface is partially mixed into subsurface waters to depths on the order of 10 m by wind waves and vertical diffusion, and is then transported with the subsurface currents. Knowledge of subsurface currents is especially important in coastal waters where flow reversals commonly occur and where valuable marine organisms may exist just below the sea surface.

Oil spreading.   Observations indicate that when oil is spilled at sea it tends to spread horizontally over the sea surface. The spreading of a surface oil slick depends on the volume and density of the oil and on environmental conditions such as wind speed, current, and wave height. Theoretical calculations of oil spreading, such as the Fay (1969) equations developed for the spreading of an instantaneous spill on a calm sea, are crude and have had only limited practical application to actual oil spills at sea.

Evaporation.   Evaporation, which removes the lighter fractions from a surface oil slick, usually causes the largest initial change in the composition of an oil spill at sea. Wind speed, spill size, the length of time that the spill has been exposed to the atmosphere, and the type of oil spilled all affect the rate of evaporation of the oil spill. High winds, a large slick, and a short exposure time all favor the evaporation process.

Emulsification.   A short time after an oil spill the heavier oil fractions begin to mix with sea water and become more viscous. Under these conditions a water-in-oil emulsion is formed which has been called "mousse" because of its similarity in color and consistency to the chocolate dessert of the same name. Oil in this form may persist at sea for months and may be advected to distant impact sites. A thick oil slick and an agitated sea state favor the oil emulsification process.

Oil in solution.   A small amount of the oil spill may actually pass into solution in sea water. This dissolved oil constitutes an increased ecological hazard since oil in this form can be ingested directly by marine organisms. An active sea state increases the amount of oil that will go into solution.

Photochemical oxidation. The less volatile, or heavier, hydrocarbon fractions in an oil spill are not very soluble in sea water. However, under the influence of sunlight these fractions interact with atmospheric oxygen to produce more soluble compounds. This photochemical oxidation process is most efficient on thin oil slicks (about 10^{-8} cm thick) and usually becomes important 1 or 2 days after the spill, when the volatile oil fractions have evaporated.

Aerosol formation. Water agitation from breaking waves produces bubbles at the sea surface. Under the proper conditions this process allows oil to actually leave the sea as an aerosol. Aerosol formation is usually of only minor importance to oil spill fate, but it can become an important factor in very rough seas.

Sedimentation. Oil has a tendency to attach to particles suspended in the water column, which then settle to the ocean bottom. Oil attached to organic material, silt, sand, and shell fragments will be removed from a surface slick. The sedimentation process is important in coastal waters, and especially in river plumes.

Biodegradation. Sea water contains many species of bacteria, molds, and yeast which consume hydrocarbons for energy and are therefore capable of removing oil from the marine environment. Biodegradation is an active process in shallow coastal waters where such organisms are more numerous because of a plentiful nutrient supply. Biodegradation slows down at lower temperatures and at depths below 1000 m.

6.2. OIL SPILL MODELING

Concern for the ecological effects and the economic impact of oil spills has led both government agencies and private industry to develop programs to prepare for, respond to, and recover from major oil spills. A key aspect of these programs is the use of oil spill models to track and predict the fate of the oil as it moves through the marine environment. These models have evolved into two types. Type I models are probabilistic and are used for long-term strategic forecasts based on archived climatological data. Type II models are deterministic and are used for specific short-term tactical forecasts based on synoptic meterological and oceanographic data.

Type I models represent hypothetical spills with individual trajectories derived from climatological information on currents and winds. These models have proven very effective for environmental assessments of activities that might result in a major oil spill, such as impact statements for proposed oil lease sale areas and for oil spill contingency planning. Type I models consider large spills of unspecified size and usually include only surface

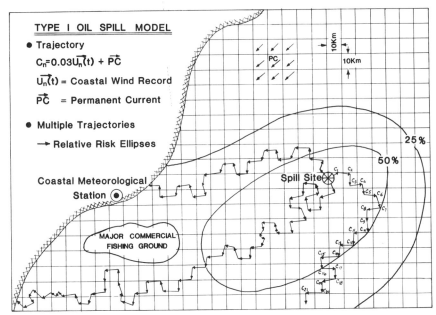

Figure 6.2. A Type I climatological oil spill trajectory model calculation, showing relative risk ellipses. Numbers on the relative risk ellipses are the result of a contour analysis of percentage "hits" in each grid area.

advection in the trajectory calculations. Shorelines, beaches, fishing grounds, and other important ecological and economic resources can easily be included as targets in Type I models. Although generally used to make climatological projections of oil spill trajectories from a hypothetical spill site, these models have also been used successfully to compute the most probable movement of actual spills when the duration of the projected trajectory is longer than the time limitations imposed by routine weather forecasts.

Climatological data are used to derive the permanent and wind-driven currents that are the two components of the Type I model advection calculation. The permanent current field is obtained from available current atlases or from a hydrodynamic ocean circulation model. Trajectories are calculated by adding the permanent current in a specific area to the local wind-drift current, which is obtained by the wind factor method explained in Chapter 2. In the Type I model calculation a computer simulation is used to track the centers of a number of hypothetical spills over a grid in three-hour time steps until the spill is beached or some predetermined maximum time limit, usually 2 to 6 months, is reached. Figure 6.2 illustrates such a simulation. Generally about 1000 hypothetical spills are simulated. Each simulation employs a continuous series of trihourly wind values to form the trajectory. The total trajectory calculation requires a local wind record of about 30 years'

TYPE II: FATE MODEL

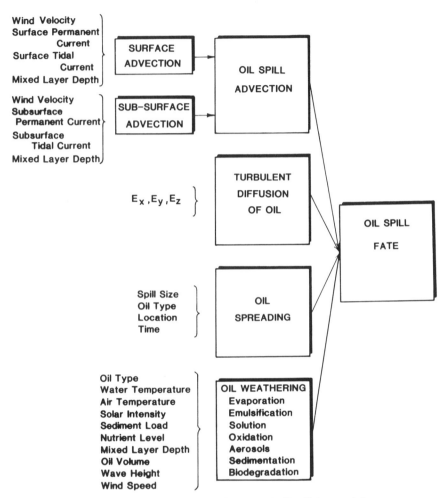

Figure 6.3. Components of a Type II oil spill fate model.

duration. Relative risk ellipses, representing the number of trajectory "hits" in a given grid area divided by the total number of trajectories initiated from the spill site, can be drawn from the results of the computer simulation.

Type II oil spill models are deterministic, in that the model input parameters are determined from synoptic data. Real-time decisions about the placement of oil spill containment or cleanup gear and research into the importance of various oil spill fate processes require the structured detail provided by a Type II model. One category of Type II model, outlined in Figure 6.3, is the fate model. In theory, this model can provide useful information concerning spill fate during an actual spill. However, the required synoptic input

TYPE II

OIL SPILL FATE AND EFFECT MODEL

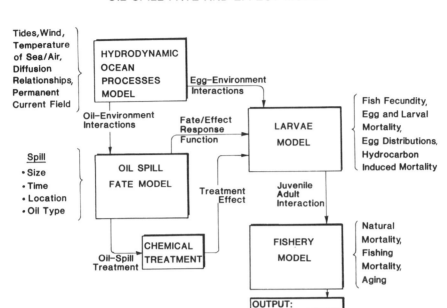

Figure 6.4. Components of a Type II oil spill fate and effect model.

data, which include oil type; size, location, and time of the spill; winds; and permanent currents, are generally not available to the modeling support group. In practice, these and other important inputs are difficult if not impossible to obtain during an actual spill. Therefore, even though a Type II fate model potentially can provide high-quality information on oil spill concentration levels, its success is limited by its dependence on input data that may not be available during an actual event.

A second category of the Type II model, the oil spill fate and effect model, is outlined in Figure 6.4. It extends the usefulness of the fate model by employing its output in an ecological effect model. Like the fate model, the fate and effect model requires detailed input data for accurate outputs and, in addition, requires information on the toxic effects of oil on marine organisms. This version of the Type II model has therefore had limited operational use, but has been employed successfully as a research tool to help evaluate the importance of various fate and effect processes.

6.3. APPLICATIONS OF TYPE I MODELS

Type I models have been used for pre-spill contingency planning and in responding to actual spills. These models are ideal for pre-spill planning

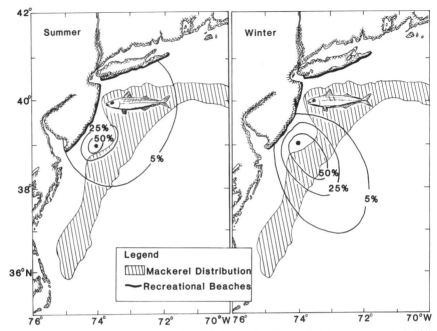

Figure 6.5. Summer and winter relative risk ellipses (see Figure 6.2) for the New York Bight overlaid on a resource chart of the mackerel distribution and Long Island/New Jersey beaches. Such charts are used in pre-spill contingency planning. (After Bishop, 1980a.)

because their climatological nature provides trajectory forecasts for spills that have not yet occurred. An example of this application, taken from Bishop (1980a), shows relative risk ellipses for hypothetical spill sites in the New York Bight. Winds used as input to the model were obtained from archived coastal wind records and the permanent current was estimated using archived density data and the coastal current model developed by Bishop and Overland (1977). Accordingly, the relative risk ellipses mirror local wind and current patterns. For example, the winter relative risk ellipses show a definite southeastward orientation, which reflects the strong northwesterly winter winds. Permanent currents in this region are weak and flow toward the southwest throughout the year, while the weak summer winds are toward the northeast. This results in summer relative risk ellipses that are more circular than the winter ellipses and are orientated in the northeast–southwest direction. The relative risk ellipses were overlaid and moved from place to place on a number of local resource charts, such as the one shown in Figure 6.5, to allow contingency planners to evaluate potential high-risk areas for spills occurring at selected sites in the planning area. For example, a major spill occurring off the coast of New Jersey would most probably affect the offshore fisheries in the winter months and the recreational beaches of Long Island and New Jersey, as well as the offshore fisheries, in the summer months. In this manner the Type I model was used to determine high-risk areas, allowing decisionmakers to position cleanup equipment and response personnel before a spill actually occurs.

Type I models have also been used to make projections for actual oil spills when the distance between the spill site and possible impact zone is large. An example of this application occurred after the June 1979 blowout of the Mexican oil well Ixtoc-1. The well caught fire and began to spill between 10,000 and 30,000 barrels of oil each day into the Gulf of Mexico until it was finally capped in March 1980. Figure 6.6 shows two satellite images of the Bay of Campeche illustrating the extent of the resulting oil slick, seen as a crescent-shaped pattern moving northeast and then northwest from the spill site. Because of the large amount of oil being discharged and the projected time needed to cap the well, U.S. government officials considered the spill a major ecological and economic threat to the coastline and decided to use available oil spill modeling techniques to determine the scope of this threat. Because of the large distance between the spill site and the coast of the United States, a Type I climatological model was considered more appropriate for the required trajectory projections than a Type II fate model. This decision was based in part on the lack of real-time synoptic data needed for a Type II model. Input to the Type I model consisted of archived wind records from coastal meterological stations and a permanent current field developed from available current atlas data. The resulting computer trajectory calculation was used to produce relative risk ellipses, shown in Figure 6.7, that indicated that trajectories initiated in June, July, and August had a small but significant probability of reaching the Texas coast with an average transit time of 30 to 60 days, and that the coastal region with the highest probability of being affected was that from Brownsville to Galveston.

The oil beached on the Texas coast, as predicted, between Brownsville and Galveston in early August 1979. Using the Type I model output, contingency planners were able to position response personnel and equipment along the most valuable recreational beaches before the spill came ashore. In addition, early fears that the resort beaches of Florida would be affected by the spill were reduced because of the model results. In short, the inexpensive and rapid computer calculation involved in a Type I oil spill trajectory model has proven useful in this and a number of applications, ranging from pre-spill contingency planning to the actual response to a major oil spill.

6.4.   APPLICATIONS OF TYPE II MODELS

Type II oil spill models can be categorized as either fate models (Figure 6.3) or fate and effect models (Figure 6.4). Fate models are designed to provide detailed information on the oil concentration distribution in time and space. One fate model developed to aid in the response to actual oil spills is the On-Scene Spill Model (OSSM) described by Galt and Torgrimson (1979). OSSM

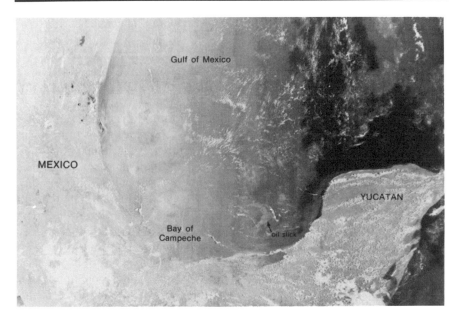

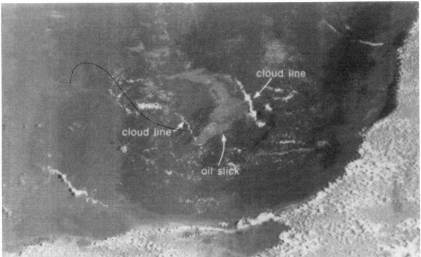

Figure 6.6. Satellite images of the Ixtoc-1 oil spill. (From U.S. Navy *Tactical Applications Guide*, 1981.)

is a computerized, interactive model which can operate at several levels of complexity, depending on user needs and data availability. The user can change model scenarios by answering questions written on a computer display board. These scenarios generally deal with variations in oil spill fate that result from changes in the input meteorological or oceanographic data. For example, if offshore winds shift to onshore, OSSM will give the resulting

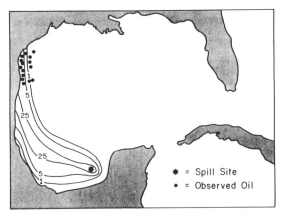

Figure 6.7. Relative risk ellipses (see Figure 6.2) and actual observations of oil concentrations from the Ixtoc-1 blowout.

variation in the amount of oil that would come ashore at a specific beach location. In practice, the application of OSSM is limited by the availability of input data. Input data generally include the size and location of the spill, present and predicted winds, permanent currents, and diffusion coefficients. The OSSM output is displayed using a computer graphics technique.

The Ixtoc-1 blowout provided an ideal test for OSSM. During this spill, a Type I model was used for long-term trajectory projections and contingency planning purposes. As the oil approached the Texas coast efforts shifted to fate modeling to provide detailed information for the short-term projections needed for day-to-day tactical purposes. Daily aerial reconnaissance to locate the oil, wind forecasts for up to 48 hours, and actual measurements of currents provided input data for OSSM. One unique application of OSSM is the use of the receptor mode, which runs the model backwards in time, to identify areas from which oil would affect specified coastal resources over a given time period. In Figure 6.8 the area enclosed by the dashed line is the largest ocean region that had to be surveyed by aircraft once a week to locate oil that might eventually come ashore on the Texas resort beaches.

The most important uses of Type II oil spill models are as a tactical aid to on-scene decisionmakers and as a research tool to gain insight into oil spill fate and effect processes. In the computer model developed by Cornillon, Spaulding, and Reed (1979) a fate model is used to develop input to an ecological effect model. The combined fate and effect model is then used to determine the dominant fate and ecological effect processes that would accompany a major oil spill near an important commercial fishing region. The general framework of this model is illustrated in Figure 6.4, showing its two components. The first is a fate component that derives input data from a hydrodynamic circulation model. The second is a fisheries model that uses information from the hydrodynamic model to make larval transport calcula-

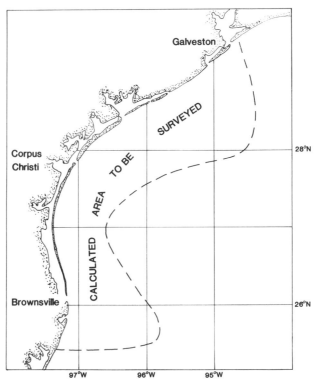

Figure 6.8. Receptor mode model run made for the Ixtoc-1 spill, showing the region that had to be surveyed by aircraft once a week to monitor for possible impact on Texas beaches. (From Galt and Torgrimson, 1979.)

tions. The fisheries model simulates fish reproduction, growth, and mortality as it would occur in nature. The fate model includes advection, diffusion, spreading, evaporation, and entrainment of the surface slick into subsurface waters. Subsurface advection is also included in the model calculations. The combination of the results of the fisheries model, which gives the distribution of fish larvae in time and space, and the results of the fate model leads to an estimate of ecological effects of the spill. The effects of chemical treatment of the oil spill are also included in the model calculation.

The Cornillon, Spaulding, and Reed fate and effect model was employed to simulate the effects of the *Argo Merchant* oil spill that resulted when this tanker broke up on Nantucket Shoals in December 1976 near the highly productive Georges Bank fishing grounds. In the first phase of the calculation the fisheries model is used to determine the annual fish catch that would occur without the spill. The ecological effect calculation is then made by combining the oil spill distribution and the fish larval distribution. In this manner, the future reduction in the annual catch due to spill-related mortality is calculated. The model indicated that the full effect of a massive spill occurs within a few years of the spill, with a maximum reduction of only about 1 percent of the annual catch. Important conclusions derived from this application of the model are that the evaporation process, which is modeled relatively accurately, is critical in reducing the initially high oil concentrations by as much as 50 percent; that the poor modeling of the shear of the surface current leads to a poor estimate of the subsurface oil distribution and of subsurface ecological effects; that surface advection is one of the most important factors in determining fisheries effects, since the most sensitive egg and larval forms are generally distributed in the near-surface waters; and finally, that the major weakness of fate and effect models is our incomplete understanding of the toxic effects of hydrocarbons on fish populations.

The main importance of oil spill fate and effect models lies in their use as a research tool to help in unraveling the complexities of the interactions of various fate and effect processes. Improvements in fate and effect models will lead to an increased interest in developing combined fate and economic impact models, which will in turn help researchers to study fate, effect, and impact in a holistic framework.

6.5. GOVERNMENT RESPONSIBILITIES

The governmental component of the oil spill pollution problem can be divided into prevention, preparedness, response, and recovery phases. Prevention is defined as government action taken to reduce the probability of a spill occurring. Certain design requirements for oil tankers are a form of prevention. For example, segregated tanks for oil and ballast water, inert gas systems to prevent explosions in tanks containing vapors, the location of water ballast tanks on the outside of the ship, dual radars, redundant steering systems, and automatic collision avoidance aids are some specific requirements proposed at international conventions to reduce the probability of oil spills at sea. A second form of prevention is government regulations used to supervise offshore drilling operations. In the United States inspectors look for deficiencies that may result in oil spillage, and in extreme cases have the authority actually to stop the drilling operation. Prevention activities are generally affected by the level of public concern about oil pollution.

The second phase of the governmental component of the oil spill pollution problem is preparedness. In the United States, the National Oil and Hazardous Substance Pollution Contingency Plan establishes a nationwide net of regional contingency plans designed to provide a coordinated response to a major oil spill. Each regional plan contains names, addresses, and telephone numbers of responsible officials; communications procedures; locations of cleanup equipment; available oil spill modeling support; and information concerning the important regional ecological and economic resources that might be damaged by an oil spill. The combination of Type I climatological oil spill models with maps of local resources, as shown in Figure 6.5, has proven to be useful in oil spill contingency planning. It may be said that preparedness bridges the gap between legislated prevention activities and the response to a major spill.

The third phase of the government's responsibility is the actual response to an oil spill. These actions are designed to minimize spill-related damage and may include oil spill modeling, containment, use of chemical agents on the spill, and planning for long-term cleanup operations. Initial responses generally concentrate on establishing the extent of the potential damage. In cases in which the spill is located at a distance from resource areas, and when sea conditions threaten loss of life or make active response measures ineffective, a passive response, with no active intervention with the spill, is appropriate. A passive response may be subject to criticism by the news media and the local population, who usually expect immediate action.

The *Argo Merchant* incident is an example of a passive response to a major oil spill. On December 15, 1976, this tanker went aground ~ 50 km southeast of Nantucket Island along the northeast coast of the United States. On December 21 the ship broke in two and an estimated 28,000 metric tons of oil spilled into the ocean. The waters near the spill site are characterized by strong, erratic tidal currents, a complicated permanent current field, and frequent severe storms. To the east of the spill site is Georges Bank, one of the world's most productive fishing grounds, and to the west are important tourist beaches, both along the mainland and on nearby Nantucket Island. The response to the spill took the form of tracking the spill by aircraft and using oil spill modeling to project the potential trajectory of the spill. Figure 6.9 illustrates the observed oil concentrations and the results of a Type I climatological oil spill model in the form of

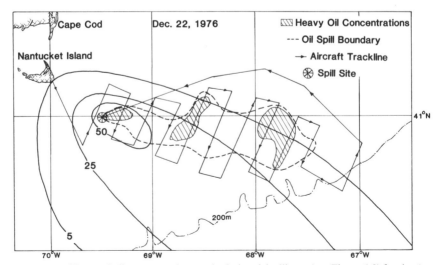

Figure 6.9.   Observed oil concentrations and relative risk ellipses (see Figure 6.2) for the *Argo Merchant* oil spill. (After Bishop, 1980b.)

relative risk ellipses. It is interesting to note how well the climatological projection agreed with the actual observations. A passive response was dictated during the *Argo Merchant* spill because environmental conditions, marked by severe storms, made mechanical cleanup impossible. Also, both observations and model projections indicated that beaches to the northwest of the spill were unlikely to be affected, and it was feared that if chemical treatment were used it might do additional harm to the major fish concentrations east of the spill site.

An active response to an oil spill could involve the use of mechanical containment and oil recovery devices and possibly the application of treating agents to the spill. Treating agents include dispersants, which are chemicals that form oil-in-water suspensions; sinking agents, which combine with the oil to create a dense mixture that will sink; burning agents, which are put on the spill to assist in its ignition and combustion; biodegradants, which promote microbial action; and sorbants, which absorb oil to form a floating mass for collection and removal. The main problem with treating agents is that they may cause additional ecological effects due to their toxicity.

The final phase of government responsibilities deals with actions which are designed to help the affected region recover from the spill. The cleanup organized by the French government after the *Amoco Cadiz* incident is a good example of a recovery operation. This supertanker contained about 220,000 metric tons of various types of oil when it broke apart 1.5 km offshore of Portsall, France on March 16, 1978. Oil spilled into the ocean and quickly turned into an emulsion. The spill caused major effects on local recreational beaches, aquaculture impoundments, and a large local fishing industry. Figure 6.10 illustrates the extent to which the coastline was covered by the spill. Strong onshore winds drove the bulk of the spilled oil into the coast and contributed to the formation of the emulsion. Shortly after the spill the emulsion was easily collected at sea and cleaned from beaches by vacuum pumping. Evaporation then caused the oil to become more tarlike, necessitating an expensive and time-consuming manual removal operation. Near Portsall the spill was a thick surface slick that was initially removed directly, but in most other regions the oil was stranded on the beach, with the highest concentrations at the high-tide line. The fate of the oil from the *Amoco Cadiz* spill is illustrated in Figure 6.11.

The recovery operation employed thousands of French government workers, military personnel, and volunteers. Weeks of manual cleaning of beaches and trucking of oil and oil-soiled materials were needed to bring the beach and local fishing grounds back from the point of complete disaster. By the first weeks of May many of the beaches appeared to be clean, although oil could still be found below the surface layers of the beaches and on the sea floor. In general, a recovery operation of this scope would only be attempted in an important resort or commercial fishing area.

The government's responsibility in the oil spill problem is to keep damage within acceptable limits. In doing this the four phrases of government ac-

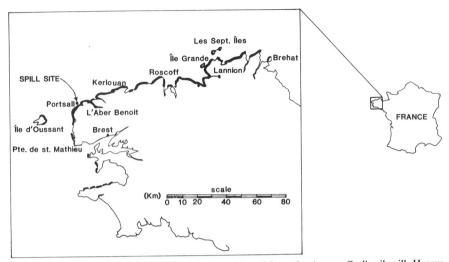

Figure 6.10.   Extent of the French coast covered by oil from the *Amoco Cadiz* oil spill. Heavy lines indicate places where oil came ashore. (After Hess, 1978.)

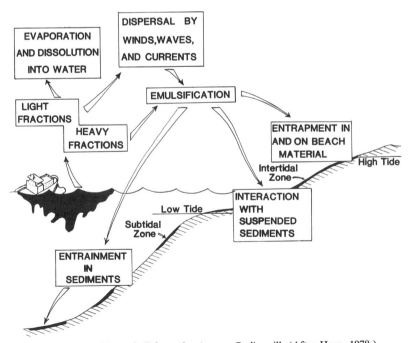

Figure 6.11.   Fate of oil from the *Amoco Cadiz* spill. (After Hess, 1978.)

tion—prevention, preparedness, response, and recovery—must make use of information concerning the fate, ecological effect, and economic impact of oil spilled at sea. Oil spillage will continue to be one of the major sources of marine pollution in future years. A more systematic application of knowledge about the ocean to the oil spill problem will lead to advances in marine protection technology, a better understanding of the long-term environmental consequences of oil spills, and more informed government regulation of this important problem in applied oceanography.

TABLE 7.1. U.S. OCEAN DISPOSAL OF WASTES, 1973–1979 (tons)[a]

Type	1973	1974	1975	1976	1977	1978	1979
Industrial wastes	5,050,800	4,579,700	3,441,900	2,733,500	1,843,800	2,548,000	2,577,000
Sewage sludge	4,808,900	5,010,000	5,039,600	5,270,900	5,134,000	5,535,000	5,932,000
Construction and demolition debris	973,700	770,400	395,900	314,600	379,000	241,000	107,000
Solid wastes	200	200	0	0	100	0	1,000
Incinerated material							
Wood	10,800	15,000	6,200	8,700	15,100	18,000	36,000
Chemicals	0	12,300	4,100	0	29,700	0	0
Total	10,844,400	10,387,600	8,887,700	8,327,700	7,401,700	8,342,000	8,653,000

Source. U.S. Environmental Protection Agency, *Ocean Dumping*, 1979 Annual Report to Congress.
[a]Includes all dumping except dredge spoil, which represents 80 to 90 percent of all materials dumped.

nated with toxic chemicals. The magnitude and toxic nature of dredge spoils makes the problem of their disposal of special importance to all users of these coastal waters.

7.2. ASSIMILATIVE CAPACITY

Factors that influence assimilative capacity include the volume of the receiving waters; the type, rate, and total amount of waste dumped; its decay time; and the sensitivities of local organisms to a given pollutant dose. Society, through government policies, generally has the option to limit ocean dumping to below the assimilative capacity. In doing this, two distinct items must be addressed. The first is the need to understand the complex physical processes that affect pollutant fate once the waste has been dumped into the ocean. Some of the physical processes that affect pollutant fate are illustrated in Figure 7.1. The second is the identification of relationships between pollutant fate, in terms of concentration levels, and the resulting ecological effects and economic impacts.

Ocean dumping of waste material is usually carried out by barge or tanker. The material is poured into a large volume of water, which dilutes it, forming a pollutant cloud. In time the cloud is advected by the large-scale flow and diffused by smaller-scale eddy processes. As the cloud increases in size the scale of motion producing advection increases, as does the eddy scale of diffusion. Eventually advection becomes dominated by the permanent circulation. At this time even large-scale eddies, some with diameters on the order of 100 km, aid in the diffusion process. Upwelling, tidal currents, and ocean frontal, mixed layer, and bottom boundary processes may also affect the waste cloud. Once an understanding of the dominant physical processes and their relationships to pollutant fate has been obtained, estimates can be made of concentration levels, ecological effects, and economic impacts of ocean dumping. In theory, government policies can then be formed to keep contamination levels within a predetermined assimilative capacity.

7.3. OCEAN DUMPING AT DWD-106

Deep Water Dumpsite 106 (DWD-106) is located ~106 nautical miles southeast of New York Harbor. It has been the site of controlled dumping of toxic wastes since the early 1970s. Prior to that it was used for the disposal of radioactive wastes, munitions, and industrial wastes. Figure 7.2 shows the location of DWD-106 in relationship to the mean positions of regional water masses and oceanic frontal zones. The site is located over the continental slope where the water depth is 1000 to 2400 m. It is an accessible dumpsite

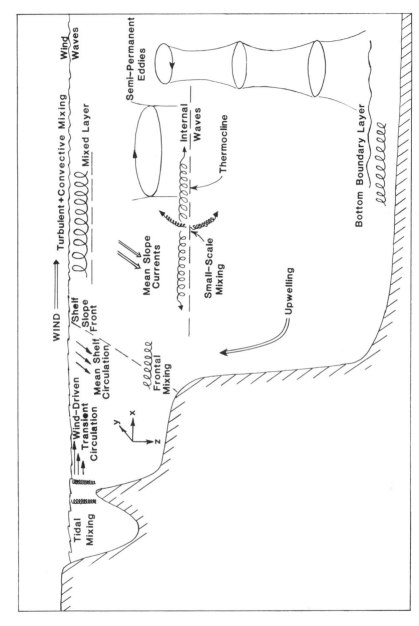

Figure 7.1. Physical oceanographic processes that influence the fate of wastes dumped in the ocean. (After Goldberg, 1979.)

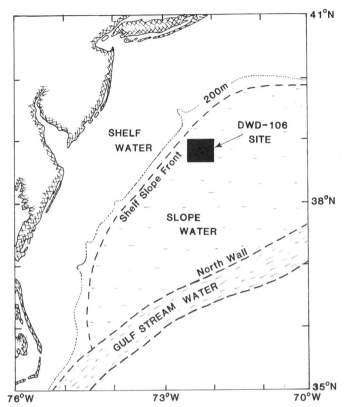

Figure 7.2.   Location of DWD-106 relative to the mean positions of the Shelf, Slope, and Gulf Stream water masses and oceanic frontal zones. (After National Oceanic and Atmospheric Administration, 1981b.)

for New York, New Jersey, and Delaware, three states with large chemical industries.

As part of the management of ocean dumping at DWD-106, the waters of the New York Bight have been studied in detail. The main objectives of these studies were to determine the effects of dumping on marine organisms and the possible economic impact on local resources. Specific project components included waste chemistry studies, dilution and diffusion studies, laboratory studies of the effects of wastes on marine organisms, monitoring of marine organisms near the dumpsite, physical oceanographic studies, and estimates of the regional assimilative capacity and its relationship to waste disposal operations. The published results of these studies (see for example, NOAA, 1981b) provide a good example of the analysis procedures involved in a major ocean dumping project. The following are brief summaries of the DWD-106 ocean dumping studies.

7.3.1. Waste Chemistry Studies

In a typical year, for example 1978, $\sim 8 \times 10^5$ m^3 of liquid industrial waste was dumped at DWD-106. Of this 4×10^5 m^3 came from the DuPont plant in Edge Moor, Delaware; 2×10^5 m^3 from the DuPont–Grasselli plant in Linden, New Jersey; 1.2×10^5 m^3 from the American Cyanamid plant, also in Linden; and the remainder from four smaller sources. The chemistry of the three major waste sources was examined in detail. Concentrations of cadmium, copper, lead, zinc, and iron were measured in the laboratory, in water samples from uncontaminated ocean areas, and in waste clouds. Since ambient concentrations were in the 2 to 500 parts per trillion range, great care was taken to avoid contamination.

The DuPont waste from the Edge Moor plant, which was derived from titanium dioxide production, was found to be a highly acidic solution rich in iron. It also contained chromium in the hundreds of parts per million (ppm) range; copper, zinc, nickel, and lead in the 10 to 100 ppm range; and cadmium in the <1 ppm range. At the rate at which the waste was dumped at DWD-106, neutralization caused the iron to precipitate as hydrous ferric oxide flocs immediately after it had been dumped into the ocean. The DuPont–Grasselli waste was found to be an alkaline solution of sodium sulfate with small amounts of dissolved metals and about 1 percent of organic carbon. When high concentrations of this material were dumped from a barge, the high alkalinity caused the precipitation of magnesium from sea water, forming magnesium hydroxide, which could be observed for up to a day after a dumping event. The American Cyanamid waste was found to be a slightly acidic solution containing ~ 2 percent organic carbon. Of the three major wastes, it was the most difficult to characterize, because it was derived from a large and diverse array of chemical processes. Its main source was the production of pesticides, water treatment chemicals, and chemicals used in the mining, paper, and rubber industries. Unlike the other major wastes, this material produced no significant particulate phase after dumping.

7.3.2. Dilution and Diffusion Studies

Waste material was barged to DWD-106 in approximately 4000-m^3 loads and was discharged by gravity from a moving barge at a discharge rate that was not to exceed a regulated maximum. During a typical year, 200 U-shaped plumes of waste, about 45 km long, were discharged within DWD-106. Initial dilution due to barge-generated turbulence and mixing with sea water was high. This method of discharge mixed the waste into a volume of water ~ 2.5 times the barge width, 3 times its draft, and as long as the dumping track. A 4000-m^3 bargeload of waste dumped over a 45-km-long track into a volume of ocean 30 m wide and 15 m deep would be diluted by a factor of $\sim$ 5000. The length of the track and the barge speed, which affects barge wake turbulence, are controllable factors in determining the initial dilution.

Several methods were used to measure dilution and diffusion of the waste plume. Drogues were deployed to track the plume. Wastes were dyed and concentrations of the dye in the waste plume were measured to estimate plume diffusion. Acoustical methods using high-frequency sound waves were used to obtain vertical cross sections of the plume if it contained particulate material. These particulate constituents of the waste plume descended slowly through the surface waters, concentrating at the thermocline, so that within the first hour or two after dumping the waste was distributed between the surface and the top of the thermocline. The affected layer therefore changed with the season, being deeper in winter than in summer. After initial dilution, the waste was mixed both horizontally and vertically by the ambient diffusion processes. Sea state, local water mass distribution, and the mixed-layer density structure all influenced this process. A more turbulent sea state led to increased diffusion, but when the sea was too rough the barge could not go to the dumpsite. In the first day or two after the release, the waste plume was found to grow from a 100 m width after about 2 hours to a width of a few hundred to a few thousand meters. As the plume width increased, the dilution was observed to increase at the same time to a final factor of approximately 10^5 to 10^6.

7.3.3. Ecological Effects Studies

Studies of the ecological effects of ocean dumping at DWD-106 involved both laboratory and field work. The ecological complexity of DWD-106 made it difficult to extrapolate the results of the laboratory studies to oceanic conditions. Field data taken during one dumping experiment were found not to be necessarily related to data taken at another time or place. Since seasonal changes in the flora and fauna of the water masses near DWD-106 are extreme, it was felt that an accurate prediction of ecological effects due to waste disposal would require synoptic field observations of the physical oceanography, a chemical analysis of the waste plume, and samples of organisms. For example, if dumping was done in an excursion of Shelf water it might affect one type of ecological system, while dumping in a warm-core eddy would affect a very different ecological system. Increasing the understanding of the relationship between waste concentrations and ecological effects was an important part of the DWD-106 ocean dumping project. The analysis of both laboratory and field data indicated no significant ecological effects at the waste concentrations that occur after waste dumping, although final conclusions were not drawn because of the complex nature of the local environmental conditions.

7.3.4. Physical Oceanographic Studies

The state of the physical environment in and around DWD-106 is a key factor governing the fate, effect, and impact of the dumped waste, and therefore constituted a major part of the ocean dumping studies. The mean positions of the three major water masses in this region are illustrated in Figure 7.2. Shelf water, extending approximately out to the 200-m bathymetric contour, exhibited large seasonal fluctuations in temperature due to seasonal heating and cooling and in salinity due to seasonal changes in freshwater runoff. The strong Shelf–Slope frontal zone separating Shelf water from the more saline Slope water mass offshore was located and charted periodically. Slope water, especially that within the upper 200 m of the water column, was also found to be affected by seasonal warming and cooling. South and east of the Slope water is the Gulf Stream, separated from the Slope water by a strong frontal zone known as the North Wall. The permanent circulation in both the Shelf and the Slope water is toward the southwest, implying a mean southwesterly trajectory for dumped waste. In addition, since DWD-106 is located offshore of the mean position of the Shelf–Slope front, the probability of waste transport toward coastal resource areas was considered to be small.

Oceanographic studies also indicated that the annual cycle of storms and solar radiation leads to a cycle of surface water stratification. From May to October a summer thermocline was observed at depths from 10 to 40 m. In addition, a permanent (winter) thermocline was found at approximately 100 to 200 m. This seasonal variation in thermal structure was found to influence the fate and effect of the dumped waste materials, since both pollutants and living organisms tend to collect in density layers, particularly above the thermocline.

Although the mean conditions shown in Figure 7.2 indicate an orderly distribution of water masses near DWD-106, on a day-to-day basis the physical oceanography is very complex, as shown in Figure 7.3. At times seaward excursions of the Shelf–Slope front bring Shelf water into the upper levels of the dumpsite, producing a complex vertical density structure. Horizontal mixing of Shelf and Slope water masses also occur across the frontal zone, usually in connection with Gulf Stream eddies, which were found near the dumpsite about 20 percent of the time. These eddies move through the Slope water on a southwesterly trajectory until they either dissipate or rejoin the Gulf Stream near Cape Hatteras. The dynamical and physical effects of eddies extend to depths of 800 to 1000 m and therefore are considered of prime importance to the ocean dumping operation. Detailed studies, such as that by Bisagni (1976), were made to determine the effects of these warm-core eddies on the ocean dumping operation. Of particular interest were eddy diameter, trajectory, forward speed, and the amount of time spent near DWD-106. As discussed in Chapters 1 and 2, the location of the Gulf Stream can be approximated by the intersection of the 15°C isotherm with the 200-m

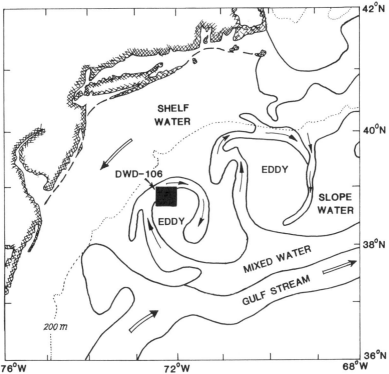

Figure 7.3. Ocean frontal analysis for July 20, 1977, showing the complexity of the DWD-106 region. (After National Oceanic and Atmospheric Administration, 1981b.)

depth. Also, according to the model for warm-core eddy formation presented in Chapter 1, the perimeters of these eddies contain Gulf Stream water. Knowing this, researchers mapped the position and extent of these eddies using remote sensing images and subsurface temperature observations. The available data indicate a mean eddy diameter of ~100 km and a mean eddy forward speed of ~ 8 km/day. Figure 7.4 illustrates a typical eddy trajectory through the Slope water. About three warm-core eddies may be expected to affect DWD-106 each year, with an average residence time within the dumpsite of~22 days. Figure 7.5 shows a vertical cross section of a Gulf Stream warm-core eddy near DWD-106.

7.3.5. Summary of DWD-106 Studies

Several important conclusions were made from the DWD-106 ocean dumping studies. First of all, after years of dumping, no significant accumulation of toxic materials was observed. Since the permanent current near DWD-106 was found to be parallel to the coast and the Shelf–Slope front was generally inshore of the dumpsite, little or no impact was expected along the

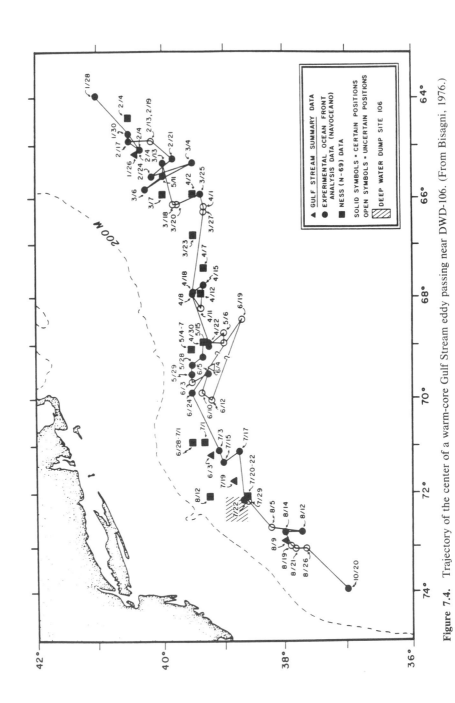

Figure 7.4. Trajectory of the center of a warm-core Gulf Stream eddy passing near DWD-106. (From Bisagni, 1976.)

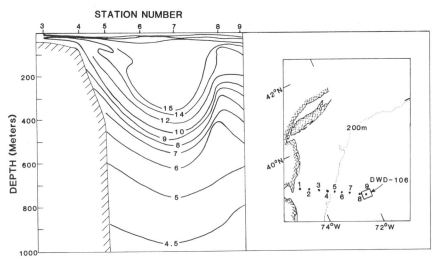

Figure 7.5. Oceanographic temperature (°C) cross section through a Gulf Stream warm-core eddy near DWD-106. (After Bisagni, 1976.)

nearby coast. It was found that an important factor in achieving rapid dilution and hence low concentrations of dumped wastes is the rate and manner of dumping. If wastes are dumped at a fairly slow rate from a rapidly moving barge, mean dilutions by a factor of 1000 to 10,000 will be reached within a few minutes. Oceanic processes will usually produce additional dilution by a factor of 10 to 100 or more within a few hours. No constituents in any of the wastes dumped are known to be harmful to marine organisms at the very low concentrations calculated or observed, although final conclusions about the ecological consequences of the waste dumping could not be made based on the study program because of the large natural variability at the dumpsite and the difficulty of collecting a sufficient number of organisms during dumping operations.

7.3.6. Assimilative Capacity at DWD-106

The concept of assimilative capacity can be used to estimate the allowable amount of waste to be dumped into a given water mass, as was done by Goldberg (1979) for dumping at DWD-106. In this study, it was assumed that the most important ecological effects from the dumping operation would be on floating plankton entrained in the waste cloud, that waste concentrations below 10 ppm (10^{-5}) would not produce any unacceptable short-term ecological effects, that concentrations below 0.1 ppm (10^{-7}) over the whole Slope water mass would not produce any unacceptable long-term ecological effects, and that the receiving water could be modeled by a closed circulation

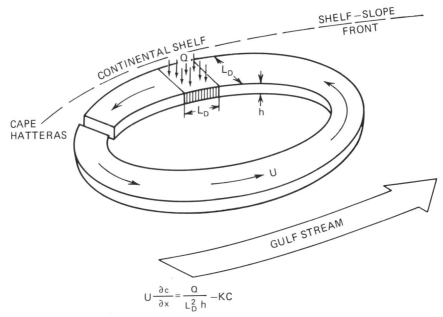

$$U\frac{\partial c}{\partial x} = \frac{Q}{L_D^2 h} - KC$$

Figure 7.6. Geometry and model equation used to estimate the assimilative capacity of the Slope water mass for dumping at DWD-106. In the model C is the pollutant concentration, U is the average Slope water current speed, and the dumpsite is h meters deep and L_D meters wide and long. The pollutant input rate is Q and its decay constant is K. (After Goldberg, 1979.)

loop~3000 km around, ~ 60 km wide, ~ 100 m deep, with a mean current of 10 cm/s.

The mathematical model that was used in this study is illustrated in Figure 7.6. The model considers waste advection and decay within a closed loop representing waste transport in the Slope water mass. A key parameter in the model is the decay time of the waste. Wastes that decay quickly need not be diluted, whereas persistent wastes may lead to a buildup in concentration levels in a closed water mass. When actual input values representing dumping at DWD-106 were used, the model output showed three distinct categories of waste concentration which depended on decay time and advection velocity. These categories, illustrated in Figure 7.7, are a "local" response for waste components with decay lives on the order of days; a "flushing" response for waste components with decay lives on the order of months, which time allows water to enter the dumpsite with no pollutant, to leave the site with a full load, and to return around the gyre with the waste completely decayed; and a "full gyre" response for waste components with decay lives on the order of a year or longer, which is about the time it takes the pollutant to become uniformly distributed around the Shelf water gyre and to reach a specific concentration level at which decay balances input.

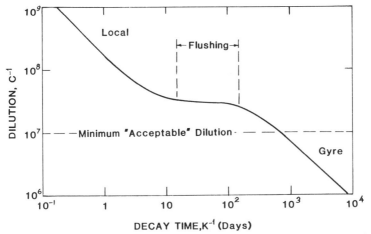

Figure 7.7. Dilution as a function of waste decay time, derived from a model calculation of assimilative capacity for DWD-106. Concentration C is the inverse of dilution and decay time (life) is the inverse of the decay constant K. (From Goldberg, 1979.)

Model calculations indicated that only the longest-lived wastes could potentially build up to concentrations of more than about 0.1 ppm (10^{-7}) throughout the full Slope water region. The calculation showed that detailed information about the decay time of the wastes, data on the long-term ecological response to various waste concentration levels, and a good description of the local physical oceanography are critical to the application of the concept of assimilative capacity to ocean dumping. A realistic application of this approach thus requires more information on these factors than was available at the time of the DWD-106 analysis. Therefore, although the concept of assimilative capacity provided some broad general guidelines for ocean dumping at DWD-106, it fell short of being the ultimate solution for this ocean dumping management problem.

7.4. ASSIMILATIVE CAPACITY AND UNCONTROLLABLE FACTORS

Some factors that alter assimilative capacity, such as accidental pollution and changes from normal oceanographic conditions, cannot be controlled. A good example of how uncontrollable factors can affect regional assimilative capacity is the Long Island beach pollution incident of June 1976. At that time, in addition to the usual ocean dumping that was being done in the area, accidental pollution and a departure from typical oceanic conditions caused the local assimilative capacity of the Shelf water mass to be exceeded. The result was beaches covered by garbage, charred wood, oil, plastics, rubber, tar, and grease. Of special concern were the tar and grease, which had a high fecal coliform count, suggesting a sewage origin.

The incident actually began in early May, when an oil spill in upper New York Bay resulted in large quantities of black oil being washed ashore along the beaches of Long Island in the form of tar balls. The Coast Guard immediately began tracking the oil and initiated cleanup operations. The oil was almost cleaned up by the end of the month, but additional events occurred which combined to worsen the situation. In the latter part of May, the flow of the Hudson River was far above normal. At that time, an oil storage tank ruptured in Jersey City and large quantities of oil spilled into the river. Shortly after that, two sewage storage tanks on southwestern Long Island exploded, spilling over two million gallons of sewage into local waters. Pier fires along the lower Hudson River at about the same time dumped tons of wreckage and debris into the water. In addition to these pollutants, a number of controlled dumpsites were in use at that time in the region of the pollution incident.

Since most of the debris was floating, a surface transport mechanism was investigated in the post-event analysis of this incident (NOAA, 1977b). Available wind and surface current meter data indicated a strong tendency for surface transport toward the Long Island beaches. Relative risk ellipses for the summer (see Figure 6.5) supported a theory that the beached material drifted out of the Hudson River and was transported by winds and currents toward the northeast.

This was not an isolated incident. Studies of this and a number of similar events point to the lack of control society has over key factors, such as accidental pollutant sources and abnormal oceanographic conditions, that may alter the regional assimilative capacity. Therefore, although this concept may be useful as a planning tool, it seems to be of only limited tactical use for marine pollution management.

Physical oceanography can be applied to the ocean dumping problem at two levels. First, for the formation of a long-term strategy for the regional assimilative capacity a detailed climatology of the regional physical oceanography must be available. This information, when combined with data on the waste chemistry and the ecological responses to various waste concentration levels, allows dumping limits to be set and regulated by government. Second, real-time information is needed for making short-term policy decisions and tactical plans to cope with pollution events that arise from departures from typical conditions. Remote sensing represents a technology that will one day allow scientists to observe the physical environment and possibly pollution itself on a near real-time basis, leading to more effective tactical application of oceanography to the ocean dumping problem.

Eight

POLLUTION MONITORING
AND ASSESSMENT

Once a pollutant has been introduced into the ocean the physical environment will dictate its fate and concentration distribution. Dose–response curves, developed under laboratory conditions, relate pollutant concentration to percentage mortality of test organisms. Therefore, in theory, given the pollutant concentration, the ecological effect of the pollutant can be determined. If a limit of acceptable damage is set at some percentage on the dose–response curve, monitoring of concentration levels or ecological effects can then be used to determine where on the curve present conditions lie. With this approach, monitoring can give the information necessary to adjust government regulations for keeping pollution within acceptable limits and can provide an early warning of potential environmental damage in time for countermeasures to be implemented.

Marine pollution monitoring may take either an event or a controlled approach. Monitoring of a major pollution event such as a large coastal oil spill or contamination of a beach usually begins after an observation of the incident. The Long Island beach pollution incident described in Chapter 7 is an example in which monitoring was initiated by an event. In the controlled approach knowledge of the levels of toxicity of specific pollutants is used to set initial waste discharge levels. Pollutant fate and the resulting ecological effects are then monitored. This controlled approach allows for revision of waste discharge levels if necessary, and has been used to monitor specific waste disposal sites, as well as large coastal ocean regions.

Figure 8.1 illustrates the relationships among the various components of the marine pollution problem viewed as a total interconnected system. The physical environment dictates pollutant fate, while ecological response func-

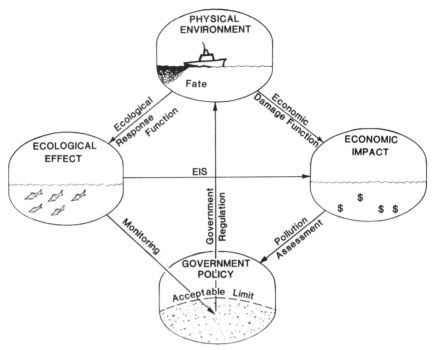

Figure 8.1. Components of a total marine pollution system. EIS is the Environmental Impact Statement.

tions, derived from dose–response curves, relate pollutant concentration levels to ecological effects on specific organisms. Similarly, economic damage functions relate pollutant fate and economic impact. This impact is measured by the monetary losses inflicted on commercially valuable economic resources such as resort beaches or fish and shellfish stocks. In addition to the direct impacts on commercially valuable marine resources, the impact of pollution on noncommercial resources such as seabirds is important. These indirect impacts are identified in the Environmental Impact Statement (EIS), which addresses the total environmental consequences of a proposed project, thereby providing the public with information on both direct and indirect environmental costs. The concerned public can then work with government representatives on policy questions before the project receives final approval. Government policymakers must assess the total benefits of the project to society and balance them against the total costs, including the indirect impacts. This process is called marine pollution assessment and represents an analysis that is completed before final approval is given to the project and, therefore, before pollution occurs.

The value of clean water to society is difficult to measure, while the benefits of potential pollution-generating activities such as offshore oil production, which results in increased employment and increased energy resources, are generally easier to put into monetary terms. The problem addressed in the marine pollution assessment process is the establishing of acceptable levels of environmental quality and pollution-producing ocean use activities. This assessment is complicated, since critical factors such as relationships between the physical environment and pollutant fate, ecological response functions, and economic damage functions are not always available. Regardless, one goal of applied oceanography is to supply the information necessary to make a reasonable assessment of the costs and benefits of various ocean use activities. An example of marine pollution assessment is presented in Chapter 5. The analysis of brine disposal into the valuable fisheries areas in the Gulf of Mexico required fate, ecological effect, and economic impact estimates to be made based on limited oceanographic data. The assessment process weighed the local environmental damage, which was found to be relatively small, against the national benefit of having a strategic oil reserve. The project was approved, and after disposal was begun monitoring indicated that damage remained within the range projected in the EIS. Another example of marine pollution assessment is the oil spill risk analysis, which is done before offshore land is leased for oil drilling. An oil spill risk model, the key component in the EIS, is critical to this marine pollution assessment problem and will be discussed later in this chapter (Section 8.4).

8.1. MONITORING A WASTE DISPOSAL SITE

The dumping of waste at designated disposal sites has become a recognized ocean use activity. These disposal sites are monitored to determine if pollutant effects are within acceptable limits and to give early warning of unfavorable trends. The choice of the parameters to be measured is based on an analysis of the most likely effects of the waste and on an understanding of those parameters that best describe these effects. A basic understanding of the physical environment is a prerequisite when monitoring a disposal site. Density stratification, mixing, and local circulation information is needed to make estimates of waste concentration levels by advection–diffusion modeling and of ecological effects from dose–response relationships. Some of this information may be obtained from the available literature, but in most cases a site designation survey is necessary to obtain more detailed data.

A number of complicated physical processes (see Figure 7.1) interact to produce an extended influence zone when waste is dumped into shallow coastal waters, as illustrated in Figure 8.2. The location and size of this zone in relationship to local ecological and economic resources determines the effects and impacts of the disposal operation. The size and exact location of the extended influence zone is determined by the type of waste dumped. For example, a solid, dense material such as concrete sinks rapidly and will

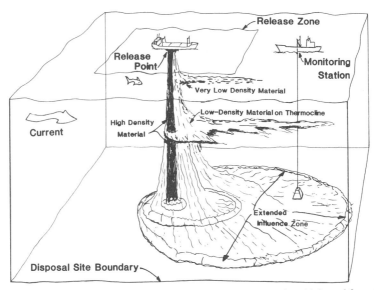

Figure 8.2. Extended influence zone for an ocean dumping operation. (Adapted from Pequegnat et al., 1981.)

come to rest approximately below the discharge point so that the extended influence zone is a vertical projection of the release zone. The physical environment will have little effect as this material passes directly to the sea floor. A fine-grained dredge material, however, produces a much larger influence zone, because advection and diffusion processes act on the material as it sinks. In cases in which the waste includes particles ranging in size from pebbles to gravel, sand, silt, and clay the extended influence zone will no longer be a simple projection of the surface release point, because each particle will sink at a rate related to its size and density. Extreme water depth and the amount of waste in liquid form also complicate the exact determination of the extended influence zone, because these factors allow oceanic transport and mixing processes to move the waste to distant locations.

The number and location of monitoring stations is determined by the size and location of the extended influence zone. Usually two to six monitoring stations are established within and near the disposal site, with more stations needed if a valuable resource is located in the area. Some of these stations are generally placed upcurrent and downcurrent of the site along the major current-flow axis. Reference, test, and control stations may be needed in regions where a polluted river flows near the disposal site. Figure 8.3 shows the relative locations of these stations. The reference station serves as the control for the test station, since any differences in the observations will represent the effects of the dumping operation. The control station is positioned so that it shows neither the effect from the polluted river nor that from

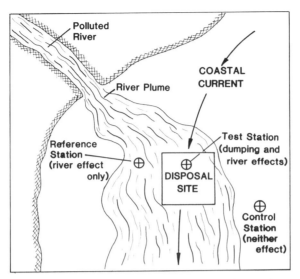

Figure 8.3. Relationship between a reference station, test station, and control station in a coastal region where a polluted river flows near a monitored ocean dumping site. (Adapted from Pequegnat et al., 1981.)

the waste dump. Comparison of data from the control, reference, and test stations will indicate whether observed effects are from natural background pollution, from the river, or from the disposal operations. These effects are determined by monitoring for the absence from the disposal site of biota characteristic of the local region; for a progressive, nonseasonal change in the composition or numbers of pelagic, demersal, and benthic organisms or in sediment or water quality; for waste material that has moved into nearby estuaries, especially productive fishery areas or beaches; and for the accumulation of contaminants in organisms near the site.

Temperature, salinity, and oxygen data, which provide information on density stratification, mixing, and local circulation, are usually measured at each monitoring station. Low values of dissolved oxygen often occur naturally in coastal waters during summer and autumn in connection with the formation of a strong seasonal thermocline. Enough data must be taken to determine whether the waste disposal operation has affected natural dissolved oxygen levels. Measurements of temperature and salinity with depth are usually made using an electronic recording instrument called an STD (for Salinity–Temperature–Depth). These data define the location of the seasonal pycnocline, a region of rapid density change, and may also help researchers to infer local circulation features; both of these factors are related to pollutant fate. Measurements of the physical environment, which are directly related to pollutant fate, effects, and impacts, are an important component of a site-specific monitoring operation.

8.2. THE MARINE ECOSYSTEMS ANALYSIS PROJECT

In the late 19th century excavation dirt and debris dumped into New York Harbor exceeded the natural capacity of the harbor to cleanse itself. As it filled shipping channels, the accumulated material began to interfere with commerce. Finally, the Army Corps of Engineers was given the responsibility to dredge the harbor and to manage the dumping of the resulting material outside the harbor entrance. Since then additional wastes have been dumped in that region, including sewage sludge, cellar dirt, construction debris, and toxic chemicals.

To monitor the effects of this site-specific ocean dumping operation the government initiated the Marine Eocsystems Analysis (MESA) New York Bight project in the early 1970s. The initial stages of the project concentrated on developing an understanding of the base-line geological, physical, chemical, and biological characteristics of the dumpsites and of the nearby ocean area. One physical oceanographic study found, for example, that water temperatures show a definite seasonal variation. In late fall and winter the water is cooled and mixed by winds into a vertically homogeneous structure, while in spring and summer solar heating causes a warm surface layer to form which is separated from the lower water by a strong thermocline. This seasonal variation in thermal structure, shown in Figure 8.4, is typical of coastal waters and has a marked influence on the advection and mixing of the dumped waste. Additional studies of the physical oceanography of the waste sites determined that the plume of sediment coming out of the Hudson River usually moves south along the coast of New Jersey and that a permanent clockwise circulation gyre occurs in the bottom waters of the New York Bight. Both of these circulation features are important to the fate of the dumped waste. Figure 8.5 shows the locations of waste dumping sites, nearby beaches, and circulation features of the inner New York Bight. Note that the dredge spoil, sewage sludge, and acid waste dumpsites are all within a few miles of each other and the nearby shoreline.

The MESA monitoring program included quarterly sediment and benthic samples over a grid of approximately 100 stations in the inner New York Bight, as shown in Figure 8.6. Chemical analysis of the sediment samples focused on the metals copper, chromium, lead, zinc, nickel, and cadmium. Especially high concentrations of all of these were found near the sewage sludge and dredge spoil sites. Cadmium levels were as high as 25 ppm near these sites, diminishing to less than 5 ppm at a radial distance of ~3 nautical miles (~5.6 km). This was of special concern because cadmium is known to concentrate in the liver and kidneys of marine organisms. MESA biological observations indicated a cadmium level of ~0.2 ppm in soft clams, which was approaching the 0.5-ppm maximum level considered acceptable for human consumption by project officials. The 0.8-ppm level found in hard clams and the 2.9-ppm found in oysters were considered above this "safe" level.

The MESA monitoring project has provided both scientists and policy-

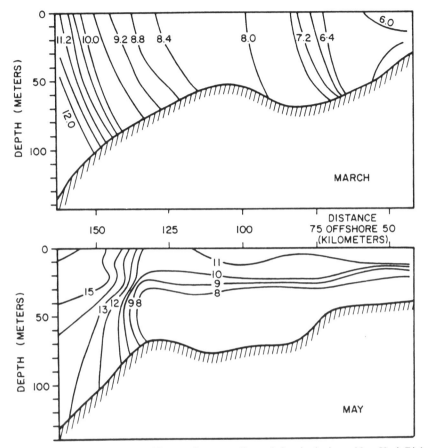

Figure 8.4. Water temperature (°C) near the waste disposal sites in the inner New York Bight during 1976. (From National Oceanic and Atmospheric Administration, 1975.)

makers with critical information on a region where there is a delicate balance between waste disposal and commercial ocean uses such as fishing, boating, and swimming. Representative contaminants and indicator organisms were selected for analysis and background physical oceanographic, geological, biological, and water quality measurements were made. This effort is an example of a site-specific monitoring project designed to provide needed information concerning pollution effects and impacts in a highly stressed marine environment.

8.3.   THE NORTHEAST MONITORING PROGRAM

In some coastal ocean regions important ocean use conflicts exist between pollution and valuable ecological and economic resources. In these regions,

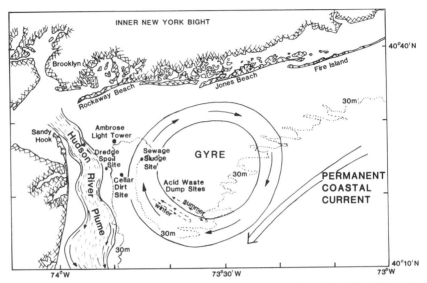

Figure 8.5. Ocean dumping sites and circulation features in the inner New York Bight (After National Oceanic and Atmospheric Administration, 1975.)

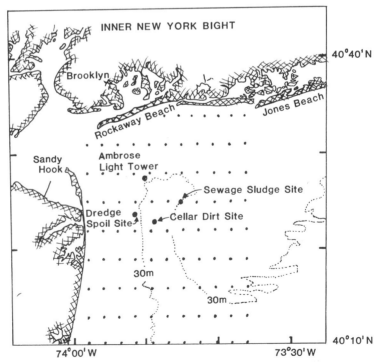

Figure 8.6. MESA monitoring stations where quarterly sediment and benthic samples were taken. (After National Oceanic and Atmospheric Administration, 1975.)

regular monitoring would provide policymakers with an early warning of unacceptable environmental damage. Government regulations could then be used to reduce deliberate waste inputs and the projected damage. The first step in the formation of a monitoring plan for a coastal ocean region is to select the monitoring stations and the parameters to be measured. Monitoring stations offshore of major estuaries are ideal for measuring specific contaminants on a regular basis from a ship. Stations should also be set around ocean dumping sites and at regular intervals over the whole continental shelf region, with special consideration given to critical habitats and areas likely to be affected by major environmental events such as a marked seasonal reduction of dissolved oxygen. The number of stations should be kept to a minimum for economy. Some general guidelines that have been developed for the types and frequency of measurements are shown in Table 8.1.

Since monitoring usually involves expensive shipboard operations, models of physical, chemical, and biological parameters are essential for the planning and implementing a cost-effective program. Models use accepted mathematical relationships to extend available observations in both time and space, and can also be used to infer one parameter, which may be difficult to measure, from other, easily measured parameters. For example, a diagnostic model for coastal currents, such as that described in Chapter 2, can be used in conjunction with density and wind observations to infer details of the coastal circulation pattern.

The Northeast Monitoring Program (NEMP) is an example of a large-scale monitoring plan. It was designed to determine the health of the marine ecosystem in the commercially valuable waters of the northeast coast of the United States. Monitoring has included both standard measurements of physical, chemical, and biological parameters, including contaminant concentration levels, and the measurement of ecological effects. Measurements have been taken at over 150 stations, with emphasis on nearshore regions affected by ocean dumping. Figure 8.7 shows the area covered and the major waste inputs for the NEMP study region. One major objective of the program was to determine the levels, trends, and variation of contamination in the water, sediments, and biota and the effects of these levels on living marine resources. Another objective was to provide in a timely manner data and relevant information to policymakers for planning and management purposes. The final objective was to determine the effects of offshore oil exploration, drilling, and ocean dumping on coastal resources, thereby providing for early warning of severe or irreversible changes in the environment caused by these activities.

The major source of pollution to the region was identified as waste input from estuaries that feed into this coastal ocean system. A wide variety of contaminants was found in estuarine sediments and biota, and changes in biota related to pollutant increases were also found. When the program started in the late 1970s, Chesapeake Bay, in which the toxic chemical

TABLE 8.1.   SUGGESTED SAMPLING FREQUENCY AND PARAMETERS TO BE MEASURED FOR A COASTAL OCEAN MONITORING PLAN[a]

Environmental Concern (Sampling Frequency)	Parameters to be Measured
Nearshore water quality (biweekly)	Dissolved oxygen Nutrients Turbidity Coliform bacteria Floatables Temperature Salinity
Oxygen depletion (four times/year)	Dissolved oxygen Turbidity Temperature Salinity Nutrients Plankton
Contaminated sediments (annually)	Cadmium Mercury Coliform bacteria
Degraded benthic community (annually)	Abundance Community structure Gill clogging

Fish and shellfish contamination (annually)	Fish
	  Cadmium
	  Mercury
	Shellfish
	  Cadmium
	  Mercury
	  Coliform bacteria
	  Pathogens

[a]After National Oceanic and Atmospheric Administration (1981a).

kepone was present; Delaware Bay, which had high levels of hydrocarbon pollution (since 70 percent of the oil delivered to the east coast passes through this Bay); and the Hudson River, which had high levels of toxic chemicals, were identified as estuaries with special problems. Pollutants from these estuaries were carried into surrounding coastal waters in dissolved and suspended form and even in the biota. Large quantities of pollutants were also introduced into the study region by direct ocean dumping. The most important disposal activities were identified as dredge spoil and sewage sludge dumping in the inner New York Bight and toxic chemical dumping at the 106-mile dumpsite. Monitoring of water quality in the NEMP focused on areas where contaminant concentrations were known or sus-

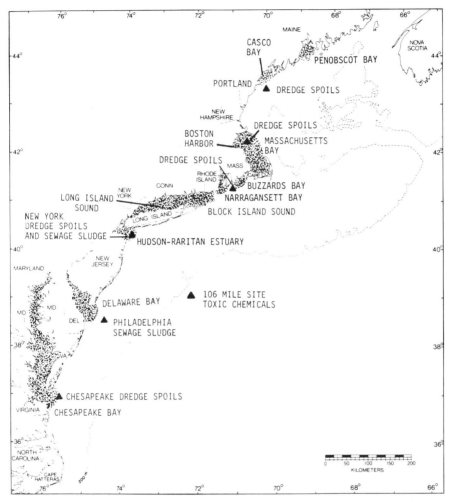

Figure 8.7. Map of the region covered by the Northeast Monitoring Program, showing the major pollutant sources as darkened areas. (After National Oceanic and Atmospheric Administration, 1981a.)

pected to be a potential problem. These areas were the New York Bight, which received a significant load of inorganic and organic dissolved and suspended nutrients from the Hudson–Raritan estuary, and the area off the Virginia Capes, which received a similar load from Chesapeake Bay.

A major conclusion of the NEMP was that many of the fisheries of the western North Atlantic are being contaminated by petroleum hydrocarbons and toxic chemicals. A number of species, from the coastal waters of the inner New York Bight to the Shelf–Slope front, showed unexpectedly high levels of contaminants. The NEMP is an example of how monitoring can be used to determine the level of pollution over a large ocean area. Measurements were used to develop bench marks for future comparisons of waste concentration levels and ecological effects. These measurements were also used to develop an overall understanding of the health of the marine ecosystem, and gave policymakers the necessary scientific information to make adjustments in the regulated levels of waste allowed to be introduced into the region.

8.4. OIL SPILL RISK ASSESSMENT

Oil spills represent the major environmental issue associated with offshore oil development in outer continental shelf (OCS) lease sale areas. Concern is strongest among those who live in coastal areas, and especially those who depend on the resources of the adjacent coastal ocean for their livelihood. The resulting ocean use conflicts have led the U.S. government to develop an oil spill risk model whose results are combined with other factors in the overall assessment of the costs and benefits of proposed oil lease sales. This assessment is made with the realization that the exact locations, number, and sizes of oil spills that might occur in connection with the proposed drilling operation cannot be determined in advance. In addition, the exact wind and current conditions that might exist during a spill cannot be known. Within these constraints a probabilistic oil spill risk model is used in this marine pollution assessment problem.

The analysis is conducted in three parts, each corresponding to a different aspect of the oil spill risk. The first part of the model considers the probability of an oil spill occurring, the second deals with spill trajectories for the times and places that these spills could occur, and the third deals with the spatial and temporal distribution of local ecological and economic resources vulnerable to the spilled oil. The results of each part of the analysis are then combined to give an estimate of the overall environmental risk associated with oil production in the lease areas. The analysis is done separately for the proposed lease areas and existing lease areas, and the results are combined to determine the cumulative or incremental risk associated with the proposed sale.

The use of this oil spill risk model for the California OCS lease sale is described in a report by Slack, Wyant, and Lanfear (1978). At the time of that study the region had a recoverable petroleum resource of an estimated 700 million barrels in addition to existing active leases that had an estimated 1300 million barrels. The expected lifetime of the drilling operation was about 20 to 25 years. In this application of the oil spill risk model statistical distributions, used to estimate the probability of a future oil spill, were developed from historical and published data. The second part of the oil spill risk model used a Type I climatological oil spill trajectory model (see Chap-

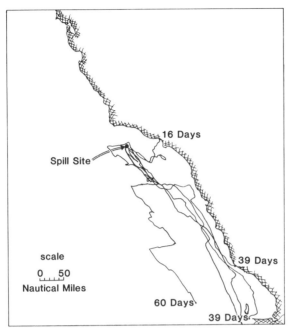

Figure 8.8. Representative oil spill trajectories for the California OCS for spring current and wind conditions. (After Slack et al., 1978.)

ter 6) to produce trajectories of hypothetical spills initiated from high-probability spill sites. Trajectories for 500 hypothetical oil spills were simulated for the summer, fall, winter, and spring seasons at each of over 70 candidate spill sites. Figure 8.8 shows a few representative trajectories for the spring season. The trajectory model produced a number of simulated spill pathways which collectively reflected both the average trend and the variability of local winds and currents. A calculation of the areal extent of each spill was added to the model by means of a simple relationship that assumed a reasonable value for the lateral diffusion coefficient. This calculation was made for various travel times and for representative spill sizes of 50, 1000, and 37,500 barrels of oil, with the assumption of a 50 percent loss of the original oil volume by evaporation. The resulting distribution of oil reaching the shoreline is shown in Figure 8.9.

The locations of 31 categories of local ecological and economic resources were digitized on the same coordinate system used for the trajectory simulations. Seasonal patterns of these resources were also considered. Typical resource charts are shown in Figure 8.10. The analysis considered the fact that the probabilities derived from the trajectory model were conditional on the spills occurring in the first place. The overall probability that oil spills

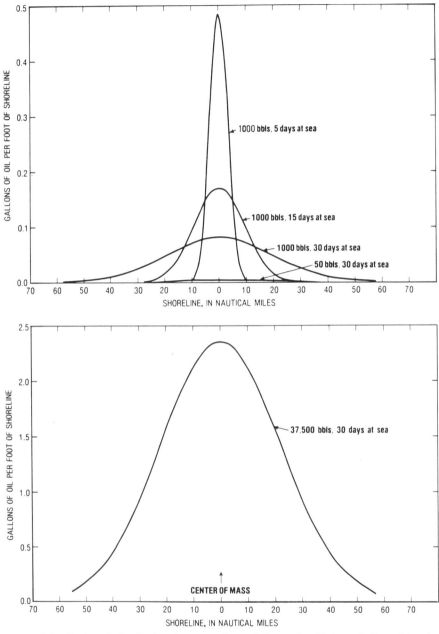

Figure 8.9. Projected distribution, in terms of travel time and spill size, of oil reaching the shoreline. (*Note:* bbls = barrels of oil.) (From Slack et al., 1978.)

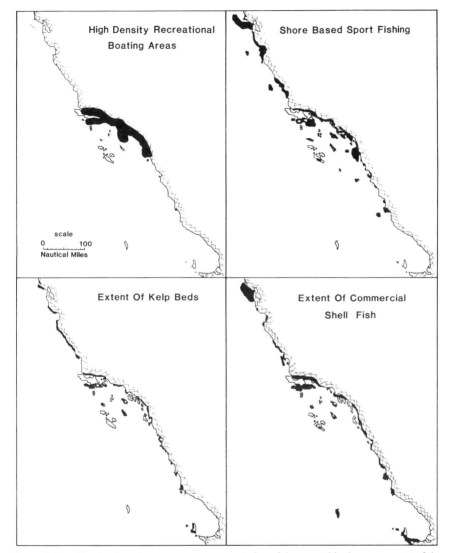

Figure 8.10.   Charts of marine resources representative of those used in the assessment of the California OCS lease sale. Dark areas are resource regions. (After Slack et al., 1978.)

will contact a particular target exactly k times during the production life of the area, $P(k)$, is therefore given by

$$P(k) = \sum_{n=k}^{\infty} P(k \mid n)\, P(n)$$

where $P(k \mid n)$ is the probability of k contacts with the resource, given the occurrence of n spills, and $P(n)$ is the probability of n spills occurring.

The results of the oil spill risk analysis model are given in Table 8.2 as percentage "hits" on the land and 31 target resources. Such information is an important part of the EIS used to identify the level of environmental damage that might occur from a proposed drilling operation. The primary reason for the development of the OCS oil spill risk model was to give government policymakers information on the amount of environmental damage that could be expected from a given site within a specific OCS lease sale area.

Model results are combined with other factors, such as the economic benefits of the drilling operation, as part of the overall marine assessment process. This assessment, which includes information derived from the oil spill risk model, will therefore favor the sale of those tracts that will not produce unacceptable environmental damage.

Model results were also used to determine the relative damage that could be expected in one OCS lease area compared with another by determining significant differences in spill risk between lease sites. One measure of this risk is the expected number of spill impacts on a resource or shoreline segment per billion barrels of oil produced and transported to shore. An example of this type of calculation is shown in Table 8.3. It gives the number of oil spills of more than 1000 barrels expected to occur and reach the shoreline during the production life of six OCS lease areas along the coasts of the United States. The second column summarizes the results of trajectory calculations using the oil spill risk model, listing the ranges of conditional probabilities of spills reaching the shore from individual production sites and transportation routes within each OCS lease area. The third column gives the expected number of spills associated with both production and transportation for the lease areas. The fourth column gives the expected number of spills reaching the shore during the production life of the six lease areas. This figure represents the sum of the products of conditional probabilities and expected numbers of occurrences of oil spills for individual tracts and transportation routes. The fifth column gives the total estimated oil production for each of the six lease areas. The last column gives the risk per unit of production expressed as the expected number of spills reaching shore per billion barrels of oil produced (calculated by dividing the fourth column by the fifth). The lowest risk exists in the Mid-Atlantic area, where the total expected number of spills reaching shore over the production life of both existing and proposed leases is only 0.19. In all other areas, the expectation of spills reaching shore is at least 6 times higher, and for Southern California, the expectation is more than 50 times higher. The main reason for the low risk values in the Mid-Atlantic area is the great distance of the tracts from the shoreline and the predominance of westerly (offshore) winds. Overall, the lease areas that stand out as being at high risk for shoreline impact per unit production are Southern California, the Gulf of Mexico, and the South Atlantic, while the Gulf of Alaska and the North and Mid-Atlantic have lower risks. Table 8.4 compares these OCS lease areas on the basis of the oil

TABLE 8.2. PERCENTAGE PROBABILITIES OF ONE OR MORE SPILLS
GREATER THAN 1,000 BARRELS OCCURRING AND HITTING RESOURCE
CATEGORIES IN THE PROPOSED SOUTHERN CALIFORNIA LEASE
SALE AREA[a]

Resource	Probability of "Hit"	
	Within 3 Days	Within 30 Days
Land	52	91
Tanner and Cortez Banks	37	54
Ranger Bank	1	1
Major marketfish	81	93
Commercial pelagic fish	88	94
Salmon (commercial fishing)	3	5
Albacore	8	34
Bonito	4	17
Tuna	1	8
Swordfish	13	25
Commercial shellfish	71	92
Seabirds (spring)	33	38
Seabirds (summer)	4	6
Seabirds (fall)	17	22
Seabirds (winter)	19	26
Sport fishing type I	83	95
Sport fishing type II	6	18
Salmon (sport fishing)	1	2
Kelp beds	51	87
Pinniped areas	29	58
High-use beaches	15	47
Harbors and marinas	5	17
Recreational boating	87	89
Sport fishing from shore	86	96
Rare and endangered species	36	73
Wildlife refuges	57	86
Sensitive biological areas	40	64
Seabird nesting areas	31	61
Rocky intertidal areas	50	83
Scuba diving areas	39	74
Clam beaches	6	21
National monuments	13	39

[a] After Slack et al. (1978).

spill risk to six general categories of marine resources expressed as the expected number of contacts with each resource category per billion barrels produced. The values given in Table 8.4 follow the same general pattern as that shown in Table 8.3. The lowest impact is for the Mid-Atlantic lease area, and the highest for Southern California.

This oil spill risk model demonstrates an application of environmental knowledge to the OSC assessment problem. Model results provide information about which lease areas could produce the largest environmental damage. With this information, environmental costs can be weighed against the benefits of drilling at a lease site. Therefore, this assessment process must also include information on the economic and social benefits of the project. Model results are a key component in this mandated assessment by government officials who are interested in keeping pollution levels within limits considered acceptable to society as a whole.

Monitoring and assessment represent two major facets of a holistic approach to marine pollution problems. The objective of monitoring is to determine concentration levels and ecological effects from an ongoing pollution-producing activity. Marine pollution assessment, on the other hand, represents a synthesis of information about the environmental costs and economic and social benefits of a proposed project. The assessment is made by government officials and must be completed before a project is given final approval. A prime objective of the assessment process is to determine if the level of pollution that might be produced by the project is acceptable to society. Monitoring and assessment provide policymakers with information with which to make decisions concerning pollution-producing ocean use activities. Improvements in the monitoring and assessment processes will require a more detailed knowledge of the physical environment, improvements in measurement techniques, and a more exact understanding of relationships between the physical environment and the fate, ecological effects, and economic impacts of marine pollution. With these improvements marine pollution monitoring and assessment will become areas of applied oceanography of greater usefulness to society.

MARINE RESOURCES

Marine resources can be defined as the collective benefits society derives from the sea, including living resources such as fish and nonliving resources such as ocean energy and minerals. In this section our emphasis will be on the application of oceanography to the management of living marine resources; the utilization of energy resources such as ocean thermal energy conversion and energy from ocean currents, salinity gradients, waves, and tides; and the mining of the ocean's hard mineral resources. The application of oceanography to marine resources will be discussed in the same systems framework we used for marine pollution. Therefore, each topic addressed will be considered to have in addition to a physical oceanographic aspect important ecological, economic, and governmental policy aspects.

Chapter

Nine

LIVING MARINE RESOURCES

The oceans hold a vast, renewable, but potentially exhaustible living marine resource in the form of fish and shellfish. The harmful effects of society on this resource have been limited until recent times, when marine pollution and poor fisheries management have resulted in documented cases of reduced yields in some of the major commercial fisheries. For example, a shift in beneficial ocean conditions combined with gross overfishing to cause the collapse of the large Peruvian *anchoveta* fishery, which fell from a peak catch of ~12 million metric tons in 1970 to ~2 million metric tons in 1973. The economics of supply and demand dictate that ever-increasing harvests from a constant or diminishing stock will require greater fishing effort and result in higher prices. At first the higher prices will stimulate increased fishing of the more popular species, but later there will be insufficient stock to support this increased effort. Since fish are the property of no one until they are caught, there is no incentive to conserve these living marine resources. Techniques that aid in the rapid and efficient filling of fishing boats, including improved fishing gear, fish-finding sonar, and searching for fish from aircraft, are being used more frequently. Remote sensing has also been employed to locate high-probability regions for fish. In this chapter we address the application of oceanography to the management of living marine resources; this is sometimes called fisheries oceanography. Discussions of the relationship between the physical environment and specific fish species, the use of remote sensing to find fish, and the collapse of the Peruvian *anchoveta* fishery are included.

149

9.1. THE PHYSICAL ENVIRONMENT AND LIVING MARINE RESOURCES

Features of the physical environment, especially its temperature, salinity, and circulation, are directly related to the distribution and abundance of various fish species. These relationships and their complication by the superimposed influence of society through overfishing and pollution are topics of ongoing study by fisheries oceanographers. Among the documented cases of damaging environmental conditions contributing to the failure of a major fishery is that of the Icelandic herring industry in the 1960s. Intense fishing combined with climatic-scale changes in the physical environment produced a large decline in this country's valuable herring industry. The abnormal harshness of the climate near Iceland during the 1960s was evident in the extent of the North Atlantic ice cover. In the spring of 1968, for example, the southward extent of the sea ice north of Iceland was greater than in any year since 1918. Johnson (1976) suggested that these conditions resulted from a shift in the global-scale atmospheric pressure pattern that brought abnormally strong northwesterly winds to the region. The changes in ice cover and the departure from the normal atmospheric pressure distribution are shown in Figure 9.1. Dickson and Lamb (1971) pointed out how the migration routes of the herring changed in response to these changes in the physical environment. Catches were reduced from 750,000 metric tons in the mid-1960s to 50,000 metric tons by 1970 as the fish moved beyond the range of the Icelandic fishing fleet. Iceland, which depended on this fishery

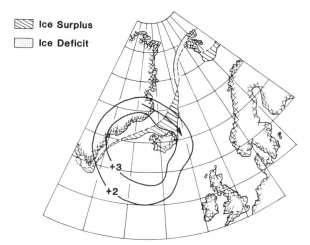

Ice Surplus
Ice Deficit

Figure 9.1. North Atlantic ice cover in 1968 compared with the long-term average. The variation from average atmospheric pressure (in millibars) is also shown for this period. The arrow indicates the abnormally strong winds that resulted from the pressure variation. (After Johnson, 1976.)

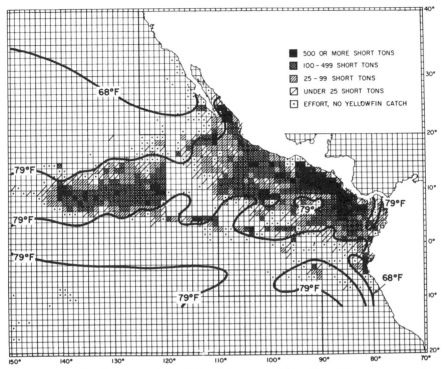

Figure 9.2. Yellowfin tuna catch for 1974 and the composite location of the mean sea surface isotherms. (After Laurs and Lynn, 1976.)

for a major portion of its economic well-being, suffered a considerable economic loss as a direct result of this environmental event.

Environmental conditions considered to be optimum for various fish species have been identified and used effectively to search for fish. The relationship between ocean temperature and commercial fish stocks has been employed to locate tuna in the eastern tropical Pacific Ocean, a region which supports a major portion of the international tuna fishery. Studies show that tuna can be found in commercial quantities in ocean regions bounded by specific sea temperatures. Each year the Inter-American Tropical Tuna Commission publishes an annual report on catch distribution. Figure 9.2 shows the distribution of the yellowfin tuna catch for 1974 and the composite positions of two limiting sea surface isotherms for the same year. The relationshp between ocean temperature and fish catch is obvious. Fisheries oceanographers have found that the most active tropical tuna fishery is actually located where the seasonal ocean temperatures remain between 79°F (26.1°C) and 84°F (28.9°C). Surface and subsurface temperature measurements, supplemented by high resolution satellite images, have

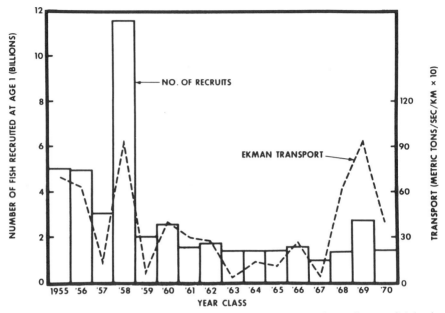

Figure 9.3. The relationship between annual Ekman transport and recruitment of Atlantic menhaden. (From Nelson et al., 1977.)

provided the data needed to aid in the successful location of tuna in this region.

The effective management of living marine resources requires an understanding of relationships between the physical environment and marine ecology. The relationships among wind-driven circulation near the edge of the Gulf Stream, larval survival, recruitment, and year class size has been studied for the Atlantic menhaden by Nelson et al. (1977). This study has revealed a strong correlation between the annual Ekman transport and recruitment (Figure 9.3). Another relationship between circulation and a commercial fishery has been derived for the Gulf of Mexico shrimp. These shrimp begin life offshore, as planktonic eggs. Larval forms must then reach estuaries which provide the environment for continued growth. Although the larvae have some mobility, the planktonic stage depends on ocean currents for onshore transport into the estuarine nursery grounds. Figure 9.4 illustrates the life cycle of the Gulf of Mexico shrimp and its relationship to favorable onshore transport. The prevailing winds in the northern Gulf of Mexico are such that fisheries oceanographers are not able to find as good a correlation between wind-driven transport and recruitment for shrimp as is found for the Atlantic menhaden. Some investigators felt that if there is a relationship between circulation and the shrimp life cycle, the Ekman model probably does not provide the best indicator of the coastal circulation field. To test this hypothesis, Leming (1982) studied the relationship between the

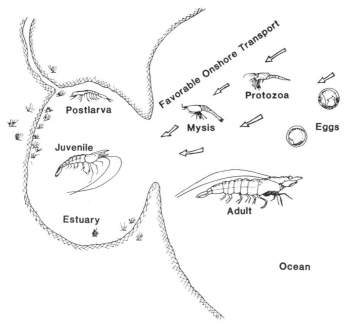

Figure 9.4.   Life cycle of the Gulf of Mexico shrimp, showing favorable onshore transport. (Courtesy of National Oceanic and Atmospheric Administration.)

brown shrimp catch and monthly coastal circulation variations using the diagnostic current model of Bishop and Overland (1977). Physical and ecological measurements were combined with model outputs in the test. Model inputs were the mean seasonal cross-shelf density gradients and mean monthly wind stress values. Model output was in terms of transport in the upper portion of the water column. Catch data were used to derive statistical relationships between model outputs and the observed shrimp stocks. Computed correlation coefficients between catch data and cross-shelf transport were much improved over those derived from an Ekman transport calculation, indicating that in the shallow coastal waters of the Gulf of Mexico both wind-driven and density-driven transport are important to larval transport and growth.

9.2.   APPLICATIONS OF REMOTE SENSING

Fisheries oceanographers have used direct measurements, ocean current models, and remote sensing techniques to derive relationships between the physical environment and living marine resources. Remote sensing, defined as the measurement of parameters by means other than direct contact, may be accomplished from satellites, aircraft, or stationary platforms. Satellite sensors, which produce synoptic images of large ocean areas, offer great promise for fisheries applications. Although the direct measurement of fish stocks from space is not presently feasible, satellite-derived information on sea surface temperatures, winds, waves, oceans fronts, and currents has been useful for locating and harvesting living marine resources. One complication with the use of remote sensing for locating fish is that the elucidation of exact relationships between the physical environment and specific species is still in the developmental stages.

The LANDSAT Menhaden Investigation was conducted jointly by the National Marine Fisheries Service (NMFS) and the National Aeronautics and Space Administration (NASA) to investigate the potential for using satellite-derived data for the prediction of fish distributions in the Gulf of Mexico. The study was conducted during the 1975 and 1976 fishing seasons and concentrated on the menhaden fishery, which is one of the oldest and most valuable in the United States. Menhaden are fished from mid-April to October during daylight hours, usually within 20 km of the shore. The basic assumption of this study, that the fish distribution would be a predictable function of the physical environment, was tested in a two-phase experiment. The first phase consisted of a series of field operations designed to derive relationships for converting remotely sensed data into fish distribution patterns. The second phase consisted of a controlled exercise performed under simulated operational conditions to demonstrate the value of the technique during actual fishing operations.

The experiment involved the coordinated use of satellite-, aircraft-, ship-, offshore-platform-, and land-based units, as illustrated in Figure 9.5. The investigation was designed to take advantage of the LANDSAT multispectral sensor system as it passed over the study area every 9 days viewing a 185-km-wide swath of ocean surface. The fishing industry provided fish distribution and abundance data from about 80 fishing vessels and numerous spotter aircraft. Relationships between LANDSAT data, fish distribution patterns, and selected oceanographic parameters (ocean color, chlorophyll, turbidity, temperature, and salinity) were hypothesized, tested, and verified. Observations were made at locations of high fish catch to establish the environmental conditions preferred by the menhaden. If fish were consistently caught in the same water conditions, then it was felt that these conditions could be used to locate the fish in the future. Three good indicators of menhaden location were water color, turbidity, and chlorophyll concentration. Since these could be inferred from the LANDSAT data, a correlation between these measurements and menhaden distributions was derived that proved to be highly significant. A spectral pattern recognition technique was used to determine by satellite the water that contained the menhaden. The LANDSAT data were digitally analyzed to identify land, water, clouds, and probable locations of menhaden schools. Values from the locations where fish were predicted were further separated into high- and low-probability categories. Figure 9.6 shows an example of the computer classification of LANDSAT data into high- and low-probability regions.

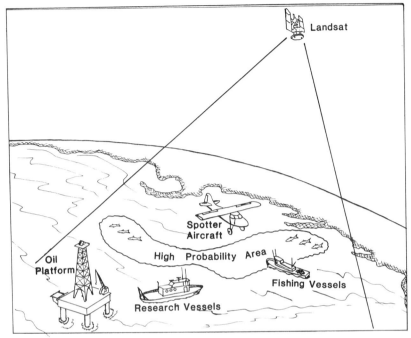

Figure 9.5. Field operations during the NMFS/NASA LANDSAT menhaden study. (After Brucks et al., 1977.)

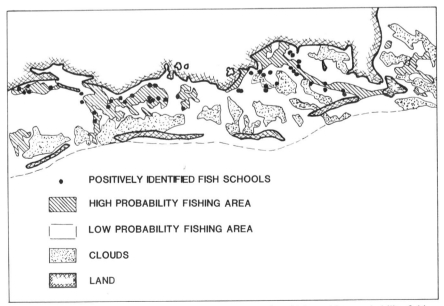

- **POSITIVELY IDENTIFIED FISH SCHOOLS**
- **HIGH PROBABILITY FISHING AREA**
- **LOW PROBABILITY FISHING AREA**
- **CLOUDS**
- **LAND**

Figure 9.6. Computer classification of LANDSAT data into high- and low-probability fishing areas during the NMFS/NASA menhaden study. (After Kemmerer and Butler, 1977.)

The LANDSAT menhaden study was considered a success, since it showed the feasibility of using satellite data to predict fish distributions on the basis of known relationships between the physical environment and fish location. Results proved that operational remote sensing could be used to reduce search time for menhaden concentrations. The data indicated that 80 percent or more of the coastal waters could be excluded from search activities. For fishing fleets this could mean a significant reduction in fuel costs and in effort exerted to achieve a profitable harvest. The quantitative relationships that were established between the menhaden distribution and oceanographic variables will also enhance future studies of fish behavior and responses to environmental changes, ultimately leading to improved management practices. This is especially important for coastal species that live and reproduce in highly dynamic and complex environments. A change in the environment could result in changes in productivity and the resultant yield. Existing management attitudes tend to relate all changes in stock size to fishing pressure. Poor yields are generally blamed on overfishing, whereas they may be due to changes in environmental conditions. Reductions in fishing pressure normally are required in either case to insure a productive fishery in the future; however, if stock reductions result from environmental changes, fishing could be allowed to return to normal levels as environmental conditions improve. The NMFS/NASA menhaden study represents a first step in an attempt to use satellite-derived information for fisheries management applications.

Another application of remote sensing to fisheries oceanography is the use of thermal-boundary charts to locate fish concentrations along the west coast of the United States. Sea surface temperature charts showing the locations of upwelling and ocean frontal regions have been produced from satellite imagery. This information is useful to fishing fleets, because these oceanic regions are areas of high fish concentrations. An example of such a chart is shown in Figure 9.7. These charts have proven particularly suited to locating albacore tuna and salmon. The analysis used to produce the charts is based on computer-enhanced data which show the details of the sea surface thermal structure. In addition to the frontal boundaries, which appear as distinct areas of dark–light contrast, smaller-scale changes in temperature are shown, represented in tones of gray. Frontal and upwelling locations are located directly on the satellite image and then transferred to a marine chart. New charts are prepared as often as cloud cover permits and are transmitted twice daily by radiofacsimile directly to the offshore fishing fleet. The analysis technique and its applications are described in detail by Breaker (1981).

Commercial fishing faces a number of problems, including more stringent regulations and increased fuel costs. Remote sensing may potentially be used to delineate areas of high fish concentrations, making commercial fishing more efficient and cost effective. With such information, fishing vessels could disperse over larger areas of high productivity and avoid overfishing small areas. Remote sensing has been used successfully in the menhaden

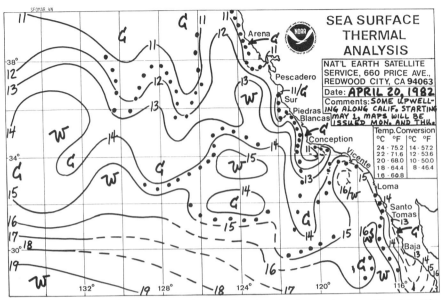

Figure 9.7. Ocean surface boundary chart used for locating temperature-sensitive fish. The dots indicate the location of surface thermal fronts and the numbers on the chart are sea surface temperature (°C). W represents warm and C represents cold water regions. (Courtesy of National Oceanic and Atmospheric Administration.)

fishery in the Gulf of Mexico and the albacore tuna and salmon fishery along the west coast of the United States. This same technology can be applied to other temperature-sensitive species. In the future, satellite-derived products will therefore become increasingly utilized to help locate and manage living marine resources.

9.3. EL NIÑO: A LIVING MARINE RESOURCE SYSTEM

The management of living marine resources requires consideration of the physical environment, ecology, economics, and government policy. One model of such a management system is illustrated in Figure 9.8. A critical component in this system is the physical environment. Optimal conditions of temperature, salinity, and circulation generally produce abundant fish stocks, whereas unfavorable environmental conditions can lower abundance or decrease availability by dispersing stocks or making them more vulnerable to predation and starvation. Society also has an effect on fish abundance through overfishing and pollution. In this management system, the government monitors fish stocks to determine their health. Although society has no control over the physical environment, government regulations can be used to control overfishing and to reduce pollution stress on local

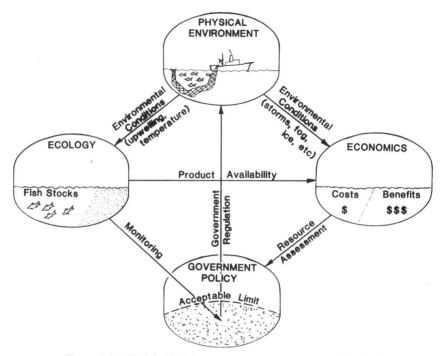

Figure 9.8.   Model of a living marine resource management system.

fisheries to an acceptable level. When environmental conditions are favorable, fish stocks are abundant and fishing effort produces large economic rewards. Therefore, the economic component of this system is linked to the ecological component through the amount of product available for harvesting. Adverse environmental conditions may necessitate exertion of a greater effort to produce a profit. Government agencies employ resource assessment and management practices to keep fishing effort and pollution within levels that will preserve the long-term welfare of the nation's fishing industry.

The *anchoveta* fishery of the Peruvian coast is a good example of the interaction of the physical environment, ecology, economics, and government policy. The coastal waters of Peru, under favorable environmental conditions, experience the ecologically productive effects of upwelling and cool, northward-flowing coastal currents. Occasionally the eastward-flowing Equatorial Countercurrent reaches the coast, spreading warm, nutrient-poor water along the coastline. This phenomenon is known in Spanish as El Niño, a reference to the Christ Child, because it generally occurs at about Christmas time. During an El Niño event the prevailing southeast trade winds that normally produce the beneficial upwelling conditions are reduced. The nutrient-poor water that is advected into the region cannot sustain the normal high level of productivity. The ecological effect of these unfavorable environmental conditions is disastrous to the local *anchoveta* stock. Under favorable conditions the *anchoveta* are abundant, and seabirds feeding on them leave massive amounts of guano (bird droppings) on nearby islands. The guano itself supports an important fertilizer industry worth millions of dollars annually for Peru. In an El Niño year the bird population and guano supply drop drastically and the coastal waters turn into a sludge of rotting plankton and fish. The *anchoveta* fishery, before the major El Niño event of 1972, supported a fishing fleet of 20,000 boats and a large number of shoreside processing plants. The economic value of Peru's fishing activities was about 350 million dollars at that time. The following year the industry all but disappeared and Peru imposed a fishing moratorium in hopes of rebuilding the *anchoveta* stock. The relationship between the physical environment (upwelling), ecology (*anchoveta* stock), and economics (fishing profits) can easily be seen in Peru's fishing industry. The final component in this resource management system is governmental policy. After a military coup in 1968 the government of Peru initiated a national fisheries management plan. The El Niño of 1972, combined with excessive fishing, had a drastic effect on the fishery. The government then took further steps to monitor and regulate fishing off the Peruvian coast, but this action came too late and fish catches continued to drop dramatically. The El Niño of 1982–1983 was considered one of the most severe on record. It resulted in at least a 50 percent decline from the already low pre-1982 catches. An extensive review of the El Niño viewed as a fisheries management problem can be found in a book by Glantz and Thompson (1981).

The use of physical oceanography in the management of living marine resources is a growing area of applied oceanography. Relationships between the physical environment and the abundance and distribution of specific fish stocks are becoming better understood. These relationships, combined with direct, simulated and remote observations of the physical environment, will help in locating and managing these valuable resources. In addition, improved techniques for monitoring and resource assessment of fish stocks will lead to more informed and effective governmental management policies. A key requisite for a living-marine-resource management system is understanding the influence of the physical environment on fish stocks. This will continue to be a prime area of research for fisheries oceanographers in the coming decades.

Chapter

Ten

RENEWABLE ENERGY RESOURCES

The world's oceans hold a vast store of nonrenewable energy resources in the form of oil, gas, and coal deposits. Because these resources are limited there is growing interest in tapping the ocean's renewable energy resources, which include temperature gradients, currents, waves, tides, and salinity gradients. These sources of energy will never completely replace fossil fuels, but may supply an increasing part of the world's energy needs, especially those of coastal nations, as we approach the next century. Extracting energy from the ocean represents an additional ocean use activity which will have to compete with fishing, oil exploration and drilling, ocean mining, navigation, and military uses for the available ocean space. This is especially true in coastal waters. To resolve the resulting ocean use conflicts will require a total systems approach which considers each resource in terms of the physical environment, ecology, economics, and government policy. Renewable energy resources are particularly advantageous because of the potential size of the resources and their nonpolluting nature. When viewed in a total systems context, however, certain disadvantages may become apparent. For example, any large-scale project, such as the building of a dam across an inlet to harness tidal energy, may have harmful consequences for the local physical and ecological environment. In addition to these problems new local, national, and even international economic issues are sure to arise as renewable ocean energy resources are developed. Ownership of the resources and the rights of owners to extract royalties are two issues to be addressed by government policymakers. The total renewable energy potential of the oceans is enormous, but it is economically recoverable only in certain areas. These include the nearshore tropical and subtropical oceans, for vertical temperature gradients; areas of strong persistent currents, such as the western boundaries of the oceans; coastal inlets, for tidal resources;

160

river outlets, for salinity gradients; and coastal areas with large persistent waves. A number of renewable energy resources will be discussed in this chapter, with emphasis on how each resource might be tapped and how knowledge of the physical environment aids in the planning of individual renewable-energy projects.

10.1. OCEAN THERMAL ENERGY CONVERSION

Over a century ago, the French physicist Jacques d'Arsonval described a technique for generating electrical power from the temperature difference between the warm surface and cold bottom waters of the tropical oceans. This process, now called Ocean Thermal Energy Conversion (OTEC), is the most promising and economically feasible renewable ocean energy resource. In the OTEC process, as illustrated in Figure 10.1, a heat engine produces electrical power from the oceanic vertical temperature difference by means of a working fluid that moves through a turbine. OTEC can work at a temperature difference of about 20°C (36°F), although efficiency increases greatly above this value. Since vertical temperature differences are greater in the tropical oceans, these regions represent the optimal OTEC resource areas. Temperature differences of at least 20°C within the upper 1000 m are found over large regions of the tropical oceans, as shown in Figure 10.2. Prime OTEC resource regions lie within 200 miles of more than 90 coastal nations and territories. A 400-megawatt (1 megawatt = 10^6 watts = 10^{13}

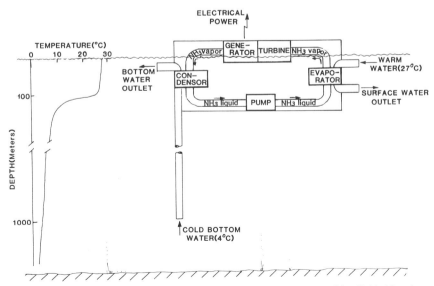

Figure 10.1.   The basic OTEC process, with ammonia (NH_3) as the working fluid. Also shown is a typical vertical temperature profile of an OTEC resource region.

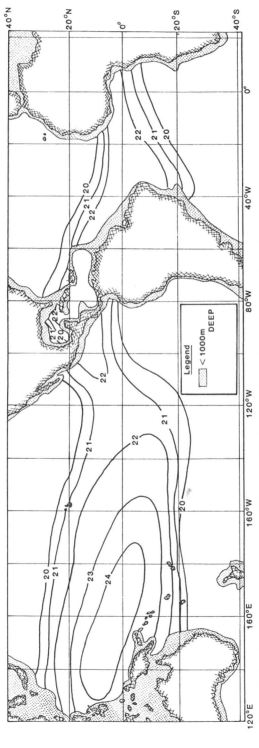

Figure 10.2. Annual average temperature differences (°C) between the surface and 1000-m depth in the Mid-Atlantic and Mid-Pacific oceans, showing the prime OTEC resource regions (those with at least a 20°C differential). (After National Oceanic and Atmospheric Administration, 1981d.)

erg/s) OTEC facility located, for example, in the Gulf of Mexico could produce enough power for approximately 60,000 homes, saving 2 million tons of coal or 6 million barrels of oil a year. This facility would require a cold-water intake pipe 30 m in diameter and 1000 m long, would use ~2.5 million gallons of ammonia or other suitable working fluid, and would require hundreds of kilometers of tubing for the heat exchanger. The continuous flow of water needed for both heating and cooling of the working fluid would equal about 20 percent of the average Mississippi River flow rate. This system would remain in the marine environment for 25 years or more, and would be subjected to fouling, corrosion, and environmental stress. In 1979 a 3-month trial of a small experimental facility off the coast of Hawaii, called mini-OTEC, was an engineering success, producing its designed requirement of 50 kilowatts (1 kilowatt = 10^3 watts) of electricity during sea trials. The mini-OTEC, illustrated in Figure 10.3, was financed by Lockheed Corporation of the United States, Dillingham Corporation of Sweden, and the state of Hawaii.

Physical oceanography and marine meteorology are prime considerations in the design, planning, and operation of an OTEC facility. The siting of the plant itself depends on locating an adequate thermal resource. The major resource areas, the tropical oceans, are regions of persistent trade winds where violent tropical cyclones sometimes form that affect large ocean areas. The winds and waves produced by these storms will drastically interfere with plant operations and therefore are a major factor in the planning of OTEC design criteria and operational procedures. Tropical cyclones occur from May to November in the Northern Hemisphere and from December to June in the Southern Hemisphere. Figure 10.4 shows the annual frequency of tropical cyclones in the Atlantic and Pacific Oceans overlaid on an OTEC resource map. Other aspects of the physical environment are also important for assessing the suitability of candidate sites for an OTEC facility. The depth of the mixed layer must be sufficient to ensure that warm water is continuously available at the water intake. This depth also influences the extent to which the plant discharge will affect the thermal structure of the mixed layer and the amount of increased biological productivity caused by OTEC-induced upwelling. The permanent circulation in the OTEC resource area will replenish the intake sea water and advect the discharge plume away from the plant. This is therefore the main factor in maintaining the thermal resource of the plant. The advection and diffusion of the OTEC thermal plume have been studied using a form of the transient plume model discussed in Chapter 5. Figure 10.5 illustrates a typical model output, showing the near-, intermediate-, and far-field regions. Model studies indicated that waters discharged above the thermocline would not sink below this region, while those discharged below the thermocline would not sink more than 50 to 100 m below the discharge point. Near-field dilution was found to be by a factor of ~10 for currents less than 50 cm/s and by a factor of ~20 for currents between 75 and 100 cm/s. Intermediate-field dilution and spreading

Figure 10.3. Mini-OTEC plant depicted as stationed off the coast of Hawaii. (Courtesy of Lockheed, Inc.)

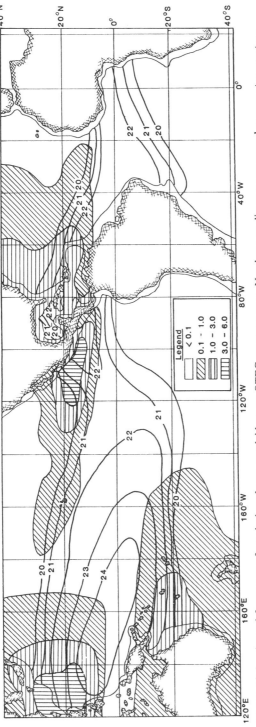

Figure 10.4. Annual frequency of tropical cyclones overlaid on an OTEC resource map. Numbers on diagram are annual average temperature difference (°C) between the surface and 1000m. (After Crutcher and Quayle, 1974.)

165

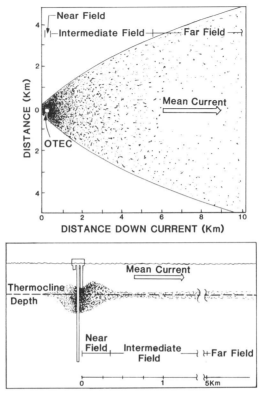

Figure 10.5. Top view and vertical cross section of a modeled thermal plume from an OTEC plant in a current of 100 cm/s. (After National Oceanic and Atmospheric Administration, 1981d.)

were also found to be governed by current speed. In areas where the current was ~10 cm/s the discharge plume was ~10 km wide and 20 m thick within 10 km downstream. A strong current of ~100 cm/s produces a narrow plume ~1 km wide 10 km downstream. An additional dilution by a factor of 10 was found within ~100 km in the far field.

A commercial OTEC operation will influence the local marine environment. Although the effects will be small compared with those from fossil fuel or nuclear power production, studies indicate that some effects may require further investigation. For example, the thermal anomaly caused by the discharge plume; the water intake operation, which will entrain organisms; the nutrient redistribution and increased biological productivity that will occur in the discharge waters; and the potential toxic hazard to organisms if the working fluid is spilled or leaked into the ocean are all effects that need further study. Although the technology required for OTEC is available, it must be developed in an ecologically safe and economically competitive manner. In addition, potential legal and political issues must be addressed by government policymakers before commercial operations commence at sea. Calculations of both the initial capital cost and the cost per kilowatt–hour imply a favorable economic future for OTEC technology. Although the first units built will cost considerably more per kilowatt–hour than fossil-fueled or nuclear power plants, expected savings due to technological improvements, assembly line production, lower operating and maintenance costs, and a long life of operation may allow OTEC power plants to become economically competitive with more conventional energy sources by the 1990s.

10.2. ENERGY FROM OCEAN CURRENTS

The sun is the energy source for the global-scale winds and the resulting ocean current systems. As discussed in Chapter 2, within each ocean basin large gyres exist in which the variation of Coriolis force with latitude produces strong western boundary currents such as the Gulf Stream. A number of methods to tap the power within these strong, relatively constant currents have been investigated. Two of the more popular ocean current energy conversion systems under consideration are the linear and the rotary systems. One representative linear system is, as shown in Figure 10.6, simply a series of parachute drogues mounted on a cable loop which moves around a drivewheel connected to an electric generator mounted on a ship. The drive cable is designed so that the parachutes open when facing an oncoming current and collapse when moving against the current, causing the drive-wheel to rotate. The loop of parachutes automatically drifts downstream

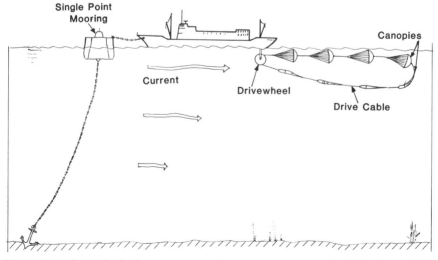

Figure 10.6.   Example of a linear ocean current energy conversion system. (After Auer, 1980.)

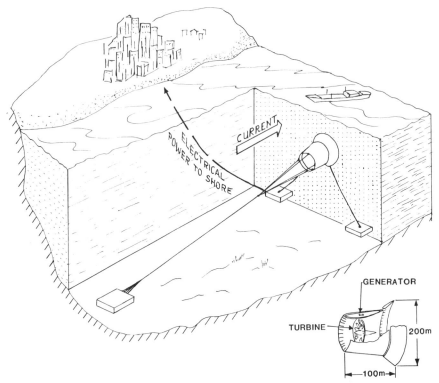

Figure 10.7.   Example of a rotary ocean current energy conversion system. (After Auer, 1980).

from the wheel so that the parachutes align themselves and the drive cable in the direction of maximum current. This linear system is relatively inexpensive and has been built and tested, demonstrating its engineering feasibility.

The Large Ocean Turbine System described by Lissaman, Radkey, Mouton, and Thompson (1979) is an example of a rotary system. It consists of a shrouded turbine approximately 100 m in length and 200 m in diameter, as shown in Figure 10.7. The shroud is designed to augment the current flow through the throat of the system. A rim-mounted counterrotating turbine is located in the throat. To optimize energy extraction, the flexible turbine blades assume the shape of a catenary as blade rotation increases. The rotational motion is converted to electricity by a series of generators driven by power take-offs positioned around the turbine rim. The system is designed to be deployed at a depth of ~30 m by a taut tripoint mooring, to reduce possible interference with surface shipping. The electricity would be transmitted to the shore by a cable.

The power that can be extracted from a current is proportional to the water transport and the square of the current speed. Water transport is usually measured in sverdrups sv; (1 sv = 10^6 m^3/s). The portion of the

Florida Current located between Miami and Bimini Island is ~80 km wide and has a transport of ~30 sv. Here the current is well channeled and steady, with speed fluctuations of only about 20 percent. This region represents an ideal location for ocean current power generation, since it is near the highly populated Florida coast. Although only a portion of this power can be captured, ocean current energy conversion could provide a valuable supplemental source of power to this region. Matrices of a number of Large Ocean Turbines, each capable of producing ~15 megawatts of power for current speeds of 1.0 to 1.5 m/s, have been proposed for deployment in the Florida Straits.

As with other forms of renewable ocean energy resources, energy production from ocean currents does not cause pollution. However, extraction of large amounts of energy from a major western boundary current such as the Gulf Stream may have some climatic-scale effects. Von Arx et al. (1974) have estimated that it is feasible to extract a maximum of about 4 percent of the total available energy from the Florida Current. They believe that if significantly larger amounts were removed the climate of Europe might be seriously affected, because the Gulf Stream is responsible for the relatively warm temperatures of Northern Europe. In addition, there could be some objection to placing large structures in a heavily used ocean area such as the Florida Straits, since the structures might pose a hazard to navigation, inhibit recreational ocean uses, or affect marine life. The cost of a large rotary ocean current energy conversion system is excessive at this time. With improvements in turbine technology and reductions in the cost of building offshore facilities, ocean current energy conversion could become a competitive energy source in highly populated regions near western boundary currents by the turn of the century.

10.3.   ENERGY FROM OCEAN WAVES

In Chapter 3 we gave a basic discussion of ocean waves. The large amounts of energy that waves possess are, if tapped, another possible source of renewable energy from the sea. According to wind wave theory, wave energy is proportional to the square of the wave height. Wave power, the rate at which energy is produced or expended, depends on both the energy in a particular wave and the speed at which the wave arrives at a given location. The wave power P_w per unit crest width for a rigid simple sinusoidal wave is given by Kinsman (1965) as

$$P_w = \frac{\rho g^2}{32\pi} H^2 T$$

where H is the wave height, T is the wave period, ρ is the water density, and g is gravity. For swell with a period of 10 s and a height of 2 m approaching an ocean wave energy device the power contained across a 10-m front L of the wave train is

$$P_wL = \frac{(1.02 \text{ g/cm}^3) \, (980 \text{ cm/s}^2)^2 \, (2 \times 10^2 \text{ cm})^2 \, (10 \text{ s}) \, (10^3 \text{ cm})}{32\pi}$$

$$P_wL \simeq 4 \times 10^{12} \text{ erg/s}$$

which is equivalent to 400 kilowatts across each 10 m of wave front.

Energy from ocean waves is potentially available nearly everywhere at sea. Highest resource regions from an economic perspective, however, are near populated regions in the major wind belts where bottom conditions tend to orient the incoming waves parallel to the coast. The west coast of the United Kingdom is an example of a prime ocean wave energy resource region. The main reason that large-scale wave energy power plants are not presently in operation is the unreliability of ocean wave energy as a constant power source, since wave heights and periods vary daily and seasonally throughout the world. Even in favorable locations it would be necessary to construct and maintain a device that could convert mechanical energy to electrical energy in an environment that at times is the most destructive known on Earth. The greatest advances in wave energy converters have come in the last few decades as new knowledge about wave propagation, interaction, and dissipation has been applied to the design of more efficient wave energy extraction devices. See a book by McCormick (1973) for an informative discussion of this subject.

Among possible wave energy extraction devices are those that use the vertical rise and fall of the water; those that use the motion of a floating body rolling, pitching, or heaving with the waves to run a turbine by means of cams or vanes; and those that focus waves in a converging channel to operate a turbine. The engineering concept of the Dam-Atoll wave energy device, developed by Lockheed Corporation, is shown in Figure 10.8. Waves enter openings near the sea surface and pass through a set of guide and turning vanes. Wave refraction causes the waves to enter the structure's center from all directions. The turning vanes cause the water to spiral downward inside the central core. This spinning water column, acting like a fluid flywheel, turns a turbine which drives a generator, thus producing electricity. Lockheed has stated that as an electrical power generator, one 80-m-diameter unit operating in waves of 7- to 10-s period could produce ~2 megawatts of power. This electricity equals the amount needed to supply approximately 300 homes in the United States. Figure 10.9 shows a series of these devices supplying power to a coastal location.

Although wave energy is generally considered to be clean and safe, several possible environmental effects have yet to be investigated and mechanisms for regulation may need to be established. Environmental issues include the effects of wave energy extraction devices on commercial and

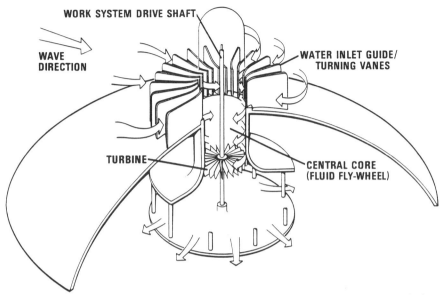

WORK SYSTEM DRIVE SHAFT

WAVE DIRECTION

WATER INLET GUIDE/ TURNING VANES

TURBINE

CENTRAL CORE (FLUID FLY-WHEEL)

Figure 10.8. The engineering concept of the Dam-Atoll wave energy conversion device. (Courtesy of Lockheed, Inc.)

recreational activities such as fishing and shipping, changes in the nutrient or temperature distribution, pollution due to accidental discharges from transportation accidents, effects on shoreline erosion, and potential visual pollution of scenic areas by overhead power cables.

10.4. ENERGY FROM OCEAN TIDES

Ocean tides, which are discussed in Chapter 4, may also be used to generate electrical power in coastal bays and estuaries that have tidal amplitudes of 5 to 15 m. A dam built across the mouth of a suitable bay would control the tidal head in the bay. The natural filling and emptying of the bay, which occur twice a day, could then be used to generate electrical power. The total tidal resource of the oceans has been estimated at $\sim 3 \times 10^6$ megawatts, of which only 2 percent is suitable for harnessing. This is a significant amount, since it represents ~ 5 percent of the 1980 worldwide power generation from all sources. The tidal power station now in operation on the Rance Estuary in France has a peak power output of 240 megawatts. It became the world's first commercial tidal power plant in 1966.

In the simplest type of tidal power plant, water is allowed to enter a single basin through open sluices which are closed at high tide. As the tide outside the basin recedes, the water is released through turbines, generating electricity. This method generates power only during a small part of the tidal cycle.

Figure 10.9. At-sea configuration of a series of Dam-Atoll wave energy devices. (Courtesy of Lockheed, Inc.)

Futhermore, the timing is dictated by the tidal cycle and does not coincide with the daily variations in demand for power. This makes integration into an available power system difficult. The situation can be improved if the power generation can be accomplished both while the basin is filling and while it is emptying. However, even in this case no power is available during periods when the height difference between the basin and the ocean is too small. If the turbines are designed so that they can also be operated as pumps, energy from other sources can be used to pump water up while the sea level difference is small. This can result in a net gain in energy, since that water can then be used to generate power when the difference in water levels between the basin and the ocean is large. This is the system in use at the Rance power plant.

The environmental effects of tidal power projects, although generally less severe than those associated with fossil fuel or nuclear power generation, must be considered. The power plant dam may affect the very tides it is built to harness, since large tides in some bays are resonance phenomena dependent on the length and depth of the bay and the period of the tidal variations. It is believed that most fish species would be able to migrate through the sluices and relatively slow-moving turbines of a tidal power plant, although the dam would be an impediment to navigation, as ships would have to pass through locks. Other environmental effects that need to be considered are the reduced vertical mixing within the basin, changed current patterns which could lead to changes in coastal erosion processes, reduced salinity, and increased pollution because of reduced mixing of the waters of the estuary with the open ocean. Despite the success of the Rance power project, development of tidal power has been slow. One of the difficulties is that the very large initial capital cost and long projected operating lifetimes of tidal power plants make cost comparisons with other power sources difficult and strongly dependent on future interest rates.

10.5. SALINITY GRADIENT ENERGY CONVERSION

Differences in salinity between freshwater rivers and nearby ocean waters provide another possible renewable ocean energy resource. Salinity gradient energy conversion would make use of the osmotic pressure difference between regions of different salt concentrations. This pressure difference can be converted into potential energy, which, in turn, can produce electricity. Osmosis is a process which selectively permits the passage of a liquid, but not of the salt which the liquid might contain in solution, through a membrane that separates two fluids. This is the same process by which plants absorb water through their roots. The principle of osmosis is illustrated in Figure 10.10, which shows a solution of high salinity separated from another of low salinity by a semipermeable membrane. This membrane allows the passage of water but not of the dissolved salts. The water flows from the region of lesser salt concentration to the region of greater salt concentration. This flow tends to equalize the salt concentrations and, in so doing, increases the volume of water on the side with the higher initial concentration. In an enclosed region the result is a difference in fluid level, which represents the osmotic pressure that can be converted into useful energy. The essential component of all ocean-salinity-gradient energy conversion devices is the semipermeable membrane, which may be a thin sheet of plastic or other material that acts as the selective separator. One technical problem connected with this process is that the membrane must be designed to withstand considerable stress from winds, waves, and currents and biological or sedimentary fouling by organisms and debris attempting to pass through it.

Salinity gradient resource areas include coastal regions just offshore of major river basins such as the Amazon, Congo, Yangtze, Ganges, and Mis-

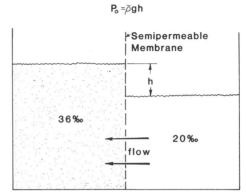

$$P_o = \bar{\rho} g h$$

Figure 10.10. Idealized diagram of the osmosis process, showing how a difference in fluid level, h, is produced by the flow of liquid across a semipermeable membrane. This height differential is equivalent to the osmotic pressure $P_o = \bar{\rho} \, g h$, in which $\bar{\rho}$ is the average density of the fluid and g is gravity. Numbers indicate salinity in parts per thousand. The osmotic pressure difference between fresh water and sea water would maintain the latter at a height of more than 200 m.

sissippi rivers. For example, it has been estimated that if the osmotic potential of one half of the mean flow of the Columbia River could be harnessed by salinity gradient coverters, $\sim 2 \times 10^6$ kilowatts of electricity could be produced (Library of Congress, 1978). This is about the same order of magnitude as the power recoverable from OTEC operations handling a similar volume of water. The technology to extract energy from salinity gradients is in its infancy. Many economic, technical, and environmental problems remain to be solved before the potential of this resource can be realized.

10.6.   ENERGY FROM OCEANIC BIOCONVERSION

Since the world's oceans offer a large potential for the collection and utilization of solar energy, proposals have been made to use the oceans for the farming of marine plants, which could in turn be converted to other useful forms of energy, such as methane gas. Initial calculations indicate that ocean farming and biomass conversion processes could produce useful quantities of hydrocarbon fuels over the reasonable production times. The best oceanic bioconversion resource areas are the nearshore regions of temperate climates, where, under favorable conditions, the net primary productivity of seaweeds may reach levels comparable to that in a tropical rain forest. Instead of developing a root system, seaweeds attach themselves to rocks or other stable objects and take nutrients directly from the sea water. Currents, especially upwelling currents, continually renew the nutrient supply. Areas of seasonal upwelling would therefore be ideal locations for ocean farming. Kelp, a prime candidate for bioconversion, flourishes from the low-tide level

to a depth of 20 to 30 m. In some places kelp beds are so dense that they are referred to as kelp forests. Ocean farming would involve more than just harvesting naturally occurring kelp beds. Kelp can be grown anywhere in the ocean if it is suitably attached to a supporting structure and supplied with nutrients pumped up from cooler, deeper water.

Although the environmental effects of a few scattered ocean farming operations would likely be minimal, if a large area of the ocean's surface were devoted to ocean farming, with cooler, nutrient-rich water being pumped to the surface, thermal anomalies of sufficient size to have a possible impact on the local weather could develop. Such effects would probably not be important in tropical regions, where the solar flux is great enough to compensate for small surface temperature differentials, but in latitudes above 30° this may be of more concern. Ocean use conflicts could arise if large areas of the ocean surface were used for ocean farming. Ships navigating through or near farming areas could not only damage the plantations, but could possibly receive damage themselves. Ship damage to the first small experimental kelp farm, near San Clemente Island off the California coast, was reported to be more of a problem than had been anticipated. On the other hand, farming the ocean surface may be more acceptable than converting a like amount of land surface to energy production.

10.7. COMBINED OCEAN ENERGY PROJECTS

Certain combinations of ocean energy projects offer very attractive prospects. For example, the combination of an OTEC plant, a desalination operation, and an ocean farming or aquaculture project has been investigated. The OTEC plant would make electricity that could be used to produce fresh water through a desalination process. Also, the cold water pumped to the surface during the OTEC operation is similar to the natural upwelling that occurs in certain highly productive ocean environments. This nutrient-rich water could be used to enhance the productivity of an ocean farming project. The three project components could be organized in such a manner that the sun and the natural ocean environment would provide a considerable range of products. Electricity, fresh water, kelp for biomass conversion, shellfish, and even finfish could be produced from such a combined operation. Figure 10.11 illustrates one possible configuration of a combined ocean energy project. Because pumping cold, deep water to the surface is an expensive operation when OTEC energy is produced alone, it is not competitive with electricity production costs from standard sources. Combining OTEC with fresh water, fish, and biomass production will make it more economically attractive, especially for remote island locations.

Renewable ocean energy resources offer a potential supplementary energy source that may reach its potential as early as the beginning of the next century. The benefits gained from the application of physical oceanography

COMBINED OCEAN ENERGY PROJECT

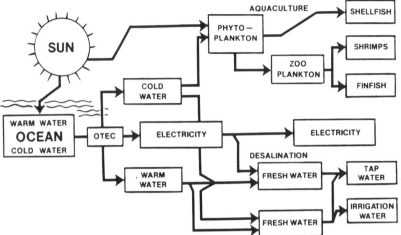

Figure 10.11.   A possible configuration for a combined OTEC, aquaculture, and desalination project. (From EDIS Magazine, 1980.)

to the development of these resources range from an increased understanding of ocean climatic conditions, which can be used for project planning and resource identification, to a better understanding of the changes in the ocean that will be produced by the project. This information will help in establishing design limitations and operating procedures. An increased understanding of physical oceanography will therefore go hand in hand with the tapping of these new energy resources. Ocean renewable energy resources are clean and offer a supplemental energy source for coastal nations. If development of the necessary technology can be matched by increased knowledge of the ocean environment in which they will operate, the maximum benefits of these new energy sources will be realized.

Eleven

MINING THE SEA FLOOR
FOR MINERAL RESOURCES

Manganese nodules were first discovered on the sea floor over a century ago, during the expedition of the *HMS Challenger*. Until the middle of this century they were considered to be of little interest other than scientific. However, when it was realized that these fist-sized nodules contained, in addition to manganese, large amounts of cobalt, copper, and nickel, commercial interest began to grow. These metals are important in the production of steel, aircraft engines, alloys, and a wide range of modern industrial materials. As the technology to extract these mineral deposits from the ocean floor develops and the world demand for the metals in the nodules increases, commercial mining for ocean mineral resources will become a reality. The issues arising from deep sea floor mining are complex problems that require an understanding of the physical environment of the mining operation, its ecological effects and economic impacts, and the governmental policy which regulates this activity. In this chapter a total systems approach will be presented that includes these four components and represents the framework within which physical oceanography is being applied to the mining of ocean mineral resources.

11.1. THE OCEAN MINING OPERATION

Although sea floor mineral resources have been found in all of the world's oceans, commercial mining interest has centered on the region of high nodule density in the equatorial North Pacific Ocean, shown in Figure 11.1. These deposits, located in water depths of 4000 to 6000 m, are of great

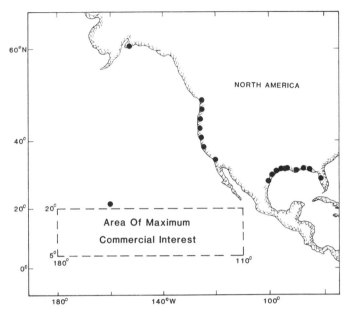

Figure 11.1. The ocean mining resource area of maximum commercial interest. The dots indicate possible processing sites. (After National Oceanic and Atmospheric Administration, 1981c.)

commercial value, because their composition is pure in the most important metals. The technology for sampling, harvesting, transporting, and extracting minerals from manganese nodules is complicated and expensive. Techniques for locating high-yield mine sites include the free-fall grab sampler and the underwater television camera. Two promising harvesting methods are a hydraulic lift system used to rake and then pump the nodules to the surface and a continuous surface-to-bottom line bucket system, which is a loop of cable with buckets attached at intervals used to hoist the nodules to the surface. The hydraulic mining system is favored by most commercial enterprises. It will collect nodules and some bottom sediments, pump the resulting slurry through a pipeline to a surface mining vessel, separate nodule materials and sediments on board, and then discharge the sediments and bottom water into the sea. Figure 11.2 shows an idealized hydraulic mining system.

During commercial operations a mining vessel will be expected to work 24 hours a day for 300 days a year. An additional 30 days is scheduled for overhauling the vessel and 35 days is scheduled for transit and downtime for weather. The nodule collector will move along the sea floor gathering materials from the top few centimeters along a 20-m-wide path. The operation will cover $\sim$ 100 km/day in closely spaced tracks, recovering $\sim$ 5000 tons of nodules. An area of up to 900 km^2 will be required to recover 1.5 million tons

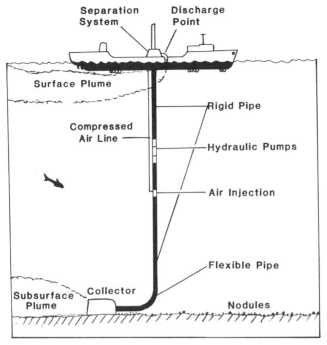

Figure 11.2.   Hydraulic ocean mining system, showing surface and subsurface waste plumes. (After National Oceanic and Atmospheric Administration, 1982.)

of nodules annually, assuming that part of the site cannot be used because of topographic features. Preliminary separation of bottom material and nodules will take place at the collector. Oversize objects will be rejected by a protective grill, while fine sediments will pass through wire-cage hoppers or be rejected hydraulically. The nodules will be drawn by a pipeline to the mining vessel. During this process a benthic plume will form following the passage of the collector. Once on the ship, the nodules and sediments will pass through a separator, and rejected materials and bottom water will be discharged into the sea, forming a surface plume.

11.2.   A SYSTEMS APPROACH TO OCEAN MINING

Ocean mining is one more activity that will be competing for the available ocean space in future decades. The application of physical oceanography to this issue should therefore be viewed in a total systems framework such as that illustrated in Figure 11.3. In this system ocean mining activities are influenced by the physical environment and may have ecological effects. Since the ocean is sometimes hostile, the physical environment may also add directly to the economic costs of the operation. Other economic costs may

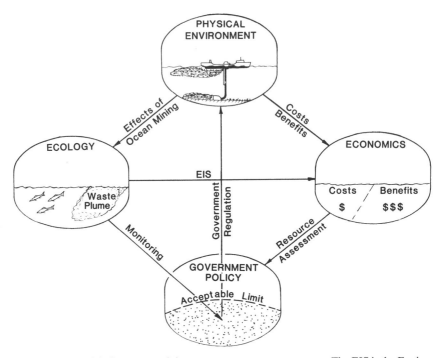

Figure 11.3. Model of an ocean mining resource management system. The EIS is the Environmental Impact Statement.

include damage done to commercially valuable marine resources in the mining area. The Environmental Impact Statement (EIS) quantifies the total ecological effect of ocean mining on both the commercial and the noncommercial aspects of the local ecology. During actual mining operations monitoring of mining discharge concentration levels and the resulting ecological effects will help to determine if environmental damages are within the level projected by the EIS. A resource assessment will be made before mining is started to determine whether the total project costs, both ecological and economic, are less than the benefits that will be derived from the project. These benefits include employment opportunities and increased availability of the scarce mineral resources. Preliminary resource assessments have indicated that seabed mining can be accomplished within acceptable limits of environmental damage, given proper monitoring and effective regulation. The final component in this systems framework, governmental policy, includes the regulation of ocean mining through licenses and permits.

11.2.1. The Physical Environment

During a commercial mining operation, the mining vessel will sweep an area of approximately 30 km by 30 km annually using a 20-m-wide collector moving at ~ 1m/s. Although some of the environmental concerns about mining are related to the movement of the collector over the sea floor, the most serious concerns relate to the sediment plume resulting from the surface discharge. Plume modeling studies indicate that the bulk of the suspended material from the collector disturbance will remain near the sea floor within about a kilometer of the collector. The surface discharge will be composed of bottom sediments, nodule fragments, and benthic biota. As this discharge enters the water some materials will sink while others will be advected and diffused within the water column. Model calculations estimate that the surface plume will be approximately 100 km long and 20 to 30 km wide and will extend to the thermocline depth of ~ 50 m.

When deep sea mining was first proposed, it was recognized that a detailed understanding of the physical environment in the mining area was critical for an adequate resource assessment to be made. In the United States this resulted in the initiation of the Deep Ocean Mining Environmental Study (DOMES), which represented the first time in history that such an extensive environmental program was conducted before a major industry was started. One objective of DOMES was to characterize the physical environment in the region of maximum commercial mining interest. Three sites within the candidate region were picked as representative of the oceanographic conditions likely to be encountered during mining. Since the greatest environmental variability in equatorial waters occurs in the poleward direction, the measurement strategy included several oceanographic stations along north–south transects at each of three sites, as illustrated in Figure 11.4. These transects crossed the two major current systems in the region, the North Equatorial Current and the Equatorial Countercurrent. Figure 11.5 is a representative set of plots showing vertical profiles of temperature, salinity, and density (sigma-*t*) during the summer and winter at the three DOMES sites. These data were used to characterize the physical environment in terms of local water masses, ocean fronts, currents, and

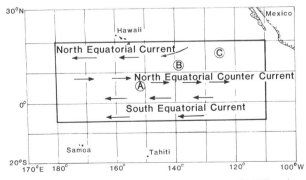

Figure 11.4. A map of oceanographic sites A, B, and C for the DOMES project, showing their positions relative to the major equatorial current systems. (From Ozturgut et al., 1978.)

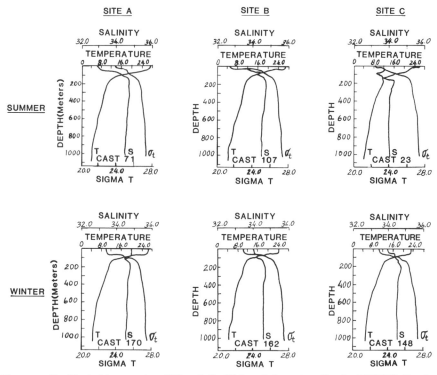

Figure .5. Typical temperature (°C), salinity (‰), and sigma-*t* profiles for DOMES sites A,
B, and   (From Ozturgut et al., 1978.)

vertical stratification. This information was useful in the plume modeling
studies because advection, diffusion, and settling rates of the surface discharge are governed by the distributions of these parameters. In addition to
physical oceanographic information, important metereological data were
also summarized. Since the mining area has a high frequency of tropical
cyclones, detailed maps of storm frequency, as shown in Figure 11.6, were
produced. These storms would drastically affect mining operations and
therefore are an important factor in developing operational plans for the
project.

11.2.2. Ecological Effects of Ocean Mining

The DOMES project addressed a number of possible ecological effects of a
commercial deep sea mining operation. For example, as the collector moves
along the sea floor the benthic biota under it will be destroyed and the
sediment plume caused by the collector will cover the nearby ocean bottom,
smothering organisms and burying the thin layer of organic material that
benthic organisms feed on. In addition, the surface discharge of the mining

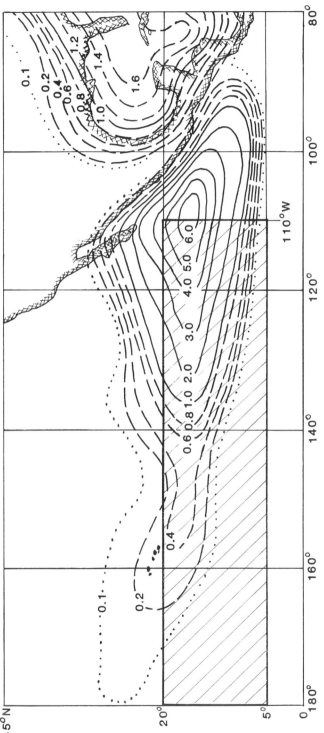

Figure 11.6. Detailed frequency chart of tropical cyclones in the primary ocean mining resource area. (After Crutcher and Quayle, 1974.)

183

ship results in increased amounts of suspended matter in the water, causing a decrease in phytoplankton photosynthesis and potentially a change in phytoplankton and zooplankton species composition. Although little is known about the life cycles of organisms that live in this deep ocean region, the DOMES project concluded that a commercial-scale ocean mining operation would have only limited effects on the local marine ecosystem. Another concern about an ocean mining operation in this region is its potential effects on the large commercial fishing industry. The Japanese and U.S. fish catch in this region amounts to approximately 40 million dollars annually. The DOMES projection of only minor ecological effects due to the ocean mining operation implies that there would not be an unacceptable impact on commercial fishing. Monitoring of the mining operation is planned to assure that this projection is accurate.

11.2.3.   The Environmental Impact Statement

A key aspect of the DOMES program was the preparation of an EIS. The EIS gave a characterization of the physical environment of the proposed mining site and information that could be used to project the ecological effects and economic impacts of the mining activity. It was also designed to aid in producing a monitoring plan, mining regulations, and in developing a resource assessment, as well as to provide information for future agreements between the United States and other nations involved with deep seabed mining. In the total systems framework illustrated in Figure 11.3 the EIS represents the link between the ecological effects and economic impacts of ocean mining. Since the EIS is a public report that identifies the harmful environmental aspects of the proposed project, it is useful to groups interested in protesting these ecological consequences even if the effects have no direct commercial value. For example, the EIS will identify the effects of ocean mining on endangered species. Special-interest groups sensitive to this issue can then take political action to influence project specifications. The EIS thus becomes a method by which the total environmental costs of the project, both direct and indirect, will be considered in the government's pre-project resource assessment of its costs and benefits. In the DOMES project the authors of the EIS concluded that only limited adverse environmental effects would occur and only in the immediate area of the mining operation. A monitoring plan was developed to insure that these projections were accurate. In addition, licensing regulations were established to guarantee that discharges from mining operations would remain within acceptable limits.

11.2.4.   Economic Impacts of Ocean Mining

Economics is a key component of the overall strategy for a proposed ocean mining operation. Considerations include the cost of the development of the technological advances needed for the operation itself, the availability of investment funding, and the ecological effects on commercial fisheries. The worldwide economic consequences of the introduction of large amounts of rare minerals are also a consideration. For example, countries in Latin America, Africa, and Asia, which are now major suppliers of copper, nickel, and cobalt, do not have the capital or technological bases to engage in ocean mining, and could possibly lose their positions as leading suppliers in the world market.

To engage in commercial-scale seabed mining, a potential operator must have a mine site with an adequate resource to sustain the venture for 20 to 25 years, sufficient knowledge of oceanic conditions at the mine site to design the mining device and operations plan, a proven method of producing a continuous flow of nodules to a surface platform, an efficient method of manufacturing products from the nodules, and methods for providing all of the support services required for nodule recovery and processing. Such support includes nodule transportation and provisions for transferring nodules between the mine-ship and nodule transport vessels, for providing the mine-ship with detailed maps of areas to be mined, for refueling, resupplying, and recrewing the mine-ship, for providing energy and other raw materials to the processing plant, and for transporting and disposing of processing wastes. In addition to possessing the technological capability to engage in commercial deep seabed mining, a potential operator must be able to obtain needed capital and legal authorization.

Deep seabed mining will lead to an increase in a nation's industrial base and create new jobs. In some nations, such as the United States, this is an important positive aspect of the effort. Each mine-ship, for example, will have a crew of about 200. A typical operation will require two mine-ships as well as an onshore processing plant that would employ about 600 people. In addition, support jobs will be created during construction and operation of a commercial mining system. Sustained long-term development of the nodule industry will depend on its economic position relative to other sources such as recycling and land mining. Technological developments, possible institutional constraints, and market conditions all affect the relative competitive position of a metals supply. The first years of nodule mining will likely be very profitable. As the industry expands into a second generation of investment and technology, possible declines in revenues may not be offset by reductions in cost.

11.2.5. Monitoring of Ocean Mining Operations

The purpose of monitoring is to determine whether the effects of a specific ocean use activity, such as ocean mining, are consistent with those predicted before the project was initiated. It represents a link between the ecological effects of the project and governmental policy, as illustrated in Figure 11.3. A goal of the DOMES project is the development of a monitoring plan that

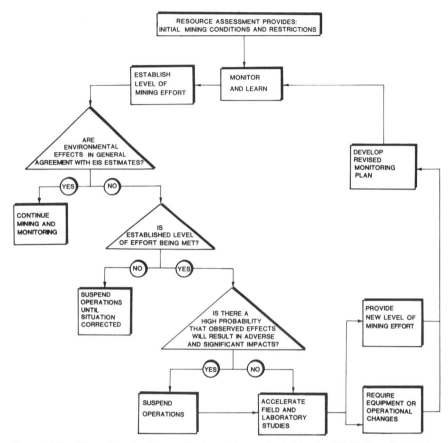

Figure 11.7. Flow chart for the monitoring strategy adopted by the DOMES project. (After National Oceanic and Atmospheric Administration, 1981c.)

would insure that unanticipated environmental effects are detected in a timely fashion. An effective monitoring plan requires the identification of the parameters to be monitored, a sampling strategy, and the use of appropriate technology to monitor these parameters in a cost-effective manner. The monitoring strategy adopted by the DOMES project is illustrated in Figure 11.7. The monitoring plan was designed such that its results could be used to adjust the level of mining operations. For example, if the observed ecological effects agree with the pre-mining projections given in the EIS, with no additional adverse effects found, monitoring will continue for about a year, after which it will be reduced in scope. If ecological effects are not consistent with those projected in the EIS, but the recommended level of operations of the mining vessels is as intended, the level of operation would be reduced. If unexpected effects are found which are not considered significant, mining would continue while research is done to determine the reason for the unanticipated findings. Finally, if unexpected ecological effects are found that are considered adverse and significant, the mining license would be suspended while decisionmakers determine the response to this unexpected situation. This monitoring approach allows project officials to regulate the mining operations as they occur.

11.2.6. Governmental Policy and Ocean Mining

The large-scale mining of the sea floor raises international economic, political, legal, and environmental issues. One important question is the ownership of ocean minerals beyond a nation's territorial waters. For living marine resources this is no problem, since fish are considered to be owned by whoever catches them. This same rule is not adopted in ocean mining. The industrialized nations, which have the greatest need for the minerals and are better able to afford the costs of developing and managing a marine mining industry, expect to reap the benefits of their ocean mining investment. On the other hand, the United Nations has affirmed that the ocean is the common heritage of all mankind, and thus the nonindustrialized developing nations expect the profits of ocean mineral resources to be shared with those who cannot afford to invest heavily in ocean mining at this time. This conflict, debated since 1973 by the United Nations Conference on the Law of the Sea, has spawned such secondary issues as the sharing of mining and extraction technology with other nations and the approval of mine sites by an International Seabed Authority. The additional issue of the extent of the economic impact of ocean mining operations on developing nations has become a focal point in the United Nations discussions. Out of these discussions have come proposals for several international regulations for diminishing the impact of deep seabed mining on developing countries. These include international control of production so that it would not interfere with land production or prices, limitation of ocean mining licenses to maintain a balance between land and sea mining, issuing of licenses for a specified amount of annual production, and compensatory payments to countries adversely affected. On the national level, in 1980 the United States enacted the Deep Seabed Hard Mineral Resource Act, which authorizes the National Oceanic and Atmospheric Administration to issue mineral licenses for ocean mining exploration and permits which authorize commercial recovery after January 1, 1988. This legislation is intended to supply sufficient regulatory certainty to enable development of a national seabed mining industry and to provide an orderly progression from no regulation to a national-level regulatory program that could serve as a model for future international regulation of ocean mining.

Ocean mining, like other topics in applied oceanography, cannot be addressed from the perspective of a single discipline such as physical oceanography alone. The issue is therefore best addressed from a total systems standpoint that includes physical oceanography, ecology, economics, and governmental policy. During the mining operation two general types of environmental disturbances will occur. At the sea floor, the benthic biota will be removed as the nodule collector passes over. Benthic organisms will also be affected by the resulting sediment plume. At the sea surface, a similar plume will be discharged by the mining vessel, causing surface ecological effects. The magnitudes of both the surface and the bottom effects will be governed by the physical environment and the level of mining activity. Significant ecological effects may lead to economic impacts on a large commercial fishery in the region. Governmental monitoring and regulation of the mining operation has been planned to keep environmental damage within acceptable limits. Ocean mining is a new and potentially important area of applied oceanography in which future technological advances must be coupled with an increased application of knowledge of the ocean environment.

MARINE TRANSPORTATION

A primary commercial and military use of the sea is transportation. Knowledge of the physical environment has been important to the masters of oceangoing vessels for centuries, because winds, waves, and currents are major factors in marine navigation. As with other categories of applied oceanography, the application of physical oceanography to marine transportation must be viewed in the context of a holistic framework that includes ecological, economic, and governmental policy aspects. For example, marine transportation is a major cause of oil pollution, which has had numerous documented effects on marine ecology. In addition, the physical environment has an economic impact on marine transportation, since harsh conditions may damage a ship or her cargo, leading to economic losses. Governmental policy aspects of marine transportation include national and international regulations concerning the safety of ships at sea. Topics in marine transportation that will be addressed in this section include optimum track shiprouting, search and rescue at sea, and military uses of the oceans.

Twelve

OPTIMUM TRACK SHIPROUTING

Environmental hazards have plagued mariners throughout history. The physical environment, in the form of waves, fog, ice, and unfavorable winds and currents, continues to affect marine transportation into modern times. The shipping industry is always looking for new ways to increase profits, minimize ship and cargo damage, and increase crew and passenger comfort by avoiding hazardous conditions. Optimum track shiprouting (OTSR) represents an application of physical oceanography to this important human and economic problem. A major step toward the development of a technique in which environmental information could be applied systematically to marine navigation was presented by James in 1957. His report described a procedure for determining the least-time track between ports which took into account winds, currents, and, most importantly, ocean waves. This procedure combined wave forecasting with empirical data on the performance of ships in various sea states. The method was tested during hundreds of voyages for the first few years after its development, with proven cost savings in the millions of dollars. The basic approach has been improved and is now being used effectively for both military and commercial applications. In this chapter discussion will focus on the use of wind, wave, and current information as an aid to marine transportation, and, specifically, OTSR.

12.1. THE EFFECTS OF WAVES ON SHIPS

On April 12, 1966, the Italian luxury liner *Michelangelo,* while steaming toward New York, was being pounded by waves reaching 20 to 25 ft (6.1 to 7.6 m) generated by a strong extratropical cyclone, as illustrated in Figure 12.1. Without warning, a huge wave towering above the others smashed

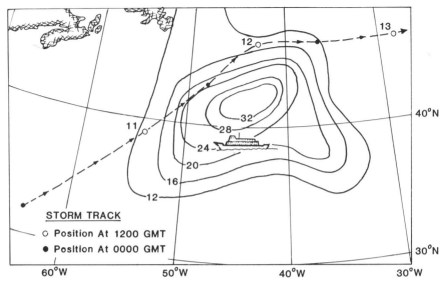

Figure 12.1. Analysis of significant wave height (in feet) for 1200 GMT (Greenwich Mean Time) April 12, 1966, showing the location of the *Michelangelo* and the storm track. (After James, 1966b.)

against the vessel, inundating her whole forward section. The steel superstructure was bent; windows on the bridge, over 80 ft (24.4 m) above the waterline, were broken; furniture flew through the air; and a bulkhead under the bridge was pushed in 10 ft (~ 3m). Three passengers were killed and 12 were injured in the brief encounter with the environment. This is but one example of the destructive effect ocean waves can have on even the most seaworthy of ships.

The effect of waves on ships is a major factor in marine transportation. In addition to their destructive capability, waves cause rolling, a side-to-side motion of the ship, and pitching, an up-and-down motion of the bow and stern. As a vessel rolls to port or starboard, the ship's inherent stability, or righting tendency, will bring her back to an upright position. The speed of the roll is an indication of this stability. A vessel is "tender" when loaded so that she is top heavy. In this condition her designed tendency to return to an upright position is reduced and her roll is slow. Such a ship has relatively poor stability. However, a ship with her load concentrated near the bottom of the vessel is "stiff" and will roll quickly from side to side. This ship has excessive stability. It is the duty of the ship's officers to ensure that their vessel is loaded so that she is neither excessively tender nor excessively stiff, since stiffness puts stress on the ship's upper structure, and a tender ship, in the extreme case, could capsize in heavy seas. A key factor in the

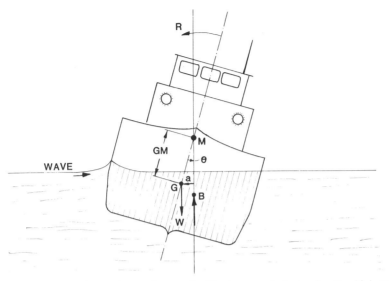

Figure 12.2.   The righting tendency of a vessel rolling to an angle θ, showing the ship's center of gravity G, its center of buoyancy B, and the metacenter M. The righting moment R is the product of the righting arm $a = GM \sin \theta$ and the weight of the vessel W.

determination of a ship's stability is the location of her center of gravity relative to her center of buoyancy. Unless the vessel's cargo shifts, the center of gravity (mass) will remain fixed. The center of buoyancy changes its location when the immersed portion of the vessel changes as she rolls. Figure 12.2 illustrates how the righting moment of a rolling vessel is related to the relative positions of the center of gravity, the center of buoyancy, and the metacenter, which is the point where the line of action of the buoyancy force intersects the ship's centerline. The righting tendency of a ship is therefore a function of the ship's stability and the angle of roll, which in turn depends on the heights and periods of the incident waves.

The wavelength of the oncoming waves is the primary factor governing the pitching motion of a ship. Waves with wavelengths less than about three-fourths of the ship's length will generally not produce serious pitching, whereas those with wavelengths equal to or greater than the ship's length will often cause severe pitching. Pitching may actually become a danger to the ship's structural integrity, since it adds extreme stress to her main fore and aft structural components. At times the ship's speed may be such that a synchronous condition occurs that results in large-amplitude pitching motions that cause pounding of the bow into oncoming seas and the possiblility of taking green seas over the bow. Excessive pounding can usually be relieved by reduction of the ship's speed or a change in course.

12.2. OPTIMUM TRACK SHIPROUTING: THE JAMES METHOD

Traditionally, the masters of oceangoing vessels have used climatological charts of winds, waves, and currents for voyage planning. As early as the mid-1800s, Lieutenant Matthew F. Maury of the U.S. Navy developed charts of ocean currents and winds based on ships' logbook data. Some of the first captains to use these charts completed round-trip voyages from New York to Rio de Janeiro in the time normally needed to make only one leg of the trip. The trip from New York to San Francisco was reduced from 6 to 4 months, and the passage from England to Australia was reduced by nearly a month. In more recent times, marine weather maps have become available directly aboard ships at sea equipped with radiofacsimile receivers. This allows day-to-day tactical decisions concerning the ship's track to be made based on present and forecasted weather conditions. OTSR is a logical extension of the mariner's use of climatological and synoptic data to improve ship efficiency.

In 1957, Dr. Richard James, an oceanographer with the U.S. Navy, published a technique that used climatological, real-time, and forecasted environmental information to calculate a least-time track between ports. He divided the problem into two basic components. The first was to determine the effects of winds, waves, and currents on ship speed. In doing this, James found that the most important factor was the effects of waves on ship performance. He then initiated a major effort to derive empirical relationships, based on logbook data, between ship speed reduction and head, beam, and following seas. A statistical analysis was done on the data and performance curves were derived for several types of ships. Three of these ship performance curves are shown in Figure 12.3; these indicate that a ship speed reduction of 10 to 20 percent may be expected for head or beam seas greater than 12 ft ($\sim$ 3.7 m) or for following seas greater than 20 ft (6.1 m). The second part of the analysis involved using the ship performance curves in conjunction with synoptic and forecasted wave conditions to calculate a least-time track between ports. Wave forecasts used in this procedure may be obtained from any of the standard wave forecasting techniques outlined in Chapter 3. James used the PNJ method.

The James OTSR procedure starts with wind forecasts covering the ocean area of the intended voyage prepared for a series of 24-hour periods. Shorter map intervals are often used near meteorological fronts and storms. These wind forecasts are derived from marine weather charts using the geostrophic approximation and applying the corrections for atmospheric instability and friction described in Chapter 3. Calculated and measured wave heights and their direction of travel are then plotted for the ocean area under analysis. Figure 12.4 shows an example of a marine weather chart and a corresponding chart of wave heights and directions. Isopleths of ship's speed are plotted over the storm area by applying the ship performance curves to the wave field. A series of 24-hour travel distances diverging from the point

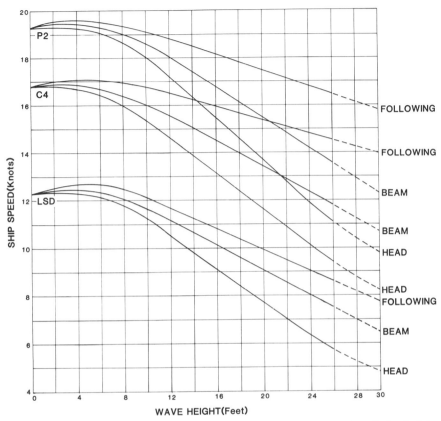

Figure 12.3.   Performance curves for three ships, as derived by James from logbook data. Dashed lines represent extrapolated values. (From James, 1957.)

where the voyage originates are now drawn on a chart which overlays the derived ship speed isopleths. This is shown in Figure 12.5. The endpoints of 1-day travel lines are joined by a smooth curve from which lines are drawn of the appropriate lengths to represent the second day's travel. The endpoints of the second day's travel lines are again joined with a smooth curve and the process is repeated for successive days to within 3 or 4 days' travel time of the destination. At this point an arc is swung from the destination to the last curve on the chart to determine the vessel's closest point of approach after several days of steaming. A final track is now added from this point to the destination to yield the total least-time voyage track. Figure 12.6 illustrates a complete calculation. The optimum voyage track is now determined based on this calculation and considering additional factors, such as the long-term weather forecast and climatological conditions of winds, waves, and currents.

The recommended track is given to the ship's master with his sailing

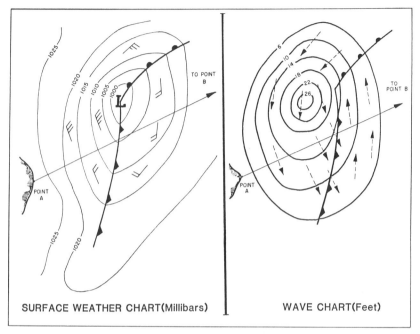

SURFACE WEATHER CHART(Millibars)   WAVE CHART(Feet)

Figure 12.4. Marine weather chart and corresponding wave chart used in an OTSR calculation for a voyage between points A and B. (From James, 1957.)

orders. During the voyage weather observations are taken by the ship's officers and oceanwide conditions are followed using radiofacsimile weather charts. Course adjustments are made as required. Since the ship's master has the final responsibility for the safety of both his crew and the cargo, OTSR guidance is generally considered advisory in nature, especially when vessel safety may be in question.

Over the years the James manual method for OTSR has proved to be superior to climatological route selection, especially for smaller ships traveling in high seas. The economic benefits of OTSR can be illustrated by using the diagram shown in Figure 12.7. Routing will add slightly to the fixed cost of a voyage, but will decrease total voyage costs, since an earlier arrival will cut fuel and crew costs and reduce weather-related damages. The time saved per voyage for a 17-knot vessel using the James manual method was calculated by Holcombe, MacDougall, and Perlroth (1959) with the assumption that the wave conditions are known. The results of their calculation are illustrated in Figure 12.8, showing the probability of gaining at least a specified number of hours (or miles) on a February North Atlantic crossing using OTSR compared with a great circle (shortest distance) route. The diagram indicates that the average gain by use of OTSR of eastbound vessels is over 2 hours, or a little more than 40 nautical miles (74 km), while west-

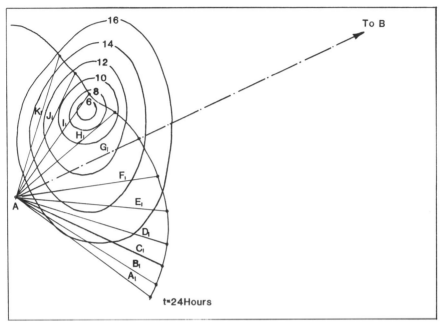

Figure 12.5. Ship speed isopleths and possible ship tracklines based on the marine weather and wave conditions shown in Figure 12.4 and on ship performance curves. In this example E_1 is the line that represents the least-time track. (From Holcombe et al., 1959.)

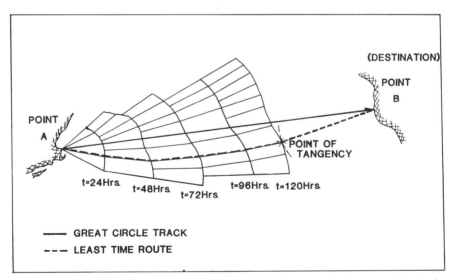

Figure 12.6. Example of a least-time track calculation. Also shown is the great circle track, which would be the shortest distance between A and B in a calm sea state. (From Holcombe et al., 1959.)

197

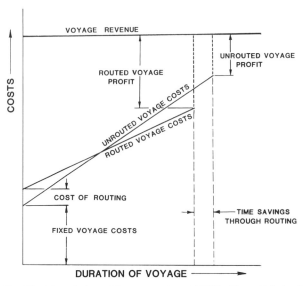

Figure 12.7. Graphical description of the economics of OTSR. (From Schukraft, 1978. Reproduced with permission of Oceanroutes Inc.)

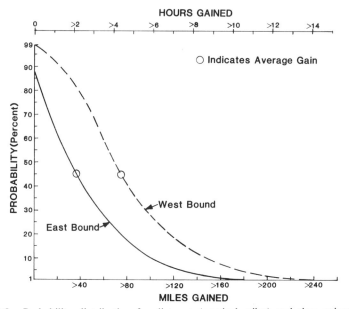

Figure 12.8. Probability distribution for distance (nautical miles) and time gained during manual OTSR compared with a great circle route in February for the North Atlantic Ocean. (From Holcombe et al., 1959.)

bound voyages save over 4 hours, or about 80 nautical miles (148 km). The diagram also indicates that approximately 10 percent of the time a minimum travel route can be chosen that will produce a saving of 6 hours eastbound and 8 hours westbound. Calculations such as this and documented cost savings have led to the use of OTSR by several of the major shipping companies of the world.

12.3. COMPUTERIZED OPTIMUM TRACK SHIPROUTING

The James method for OTSR requires subjective decisions to be made by a trained forecaster. Since the initial development of OTSR, advances in computer technology have led to the automation of the basic technique. For example, Bleick and Faulkner (1971) summarized several sophisticated techniques for OTSR using digital computer technology. As with the manual method, computerized OTSR requires knowledge of the sea state during the voyage. Computer-generated wave forecasts fit naturally into a computerized OTSR scheme. Figure 12.9 illustrates a grid used in a typical computerized OTSR scheme. The number of grid points and their geometry are determined by the size and speed of the computer used for the calculation. The calculated OTSR route will be along straight line segments of the grid which start from the ship's base location (location B in Figure 12.9) and go toward all points in the next vertical column in the ship's intended direction of travel. The point toward which the ship is sailing at a given time is called the target (shown as location T in Figure 12.9). Forecasted wave conditions and ship performance information are used to calculate a least-time track between each vertical row of points as the simulated ship travels across the ocean. An iterative process is used to determine the combination of tracks

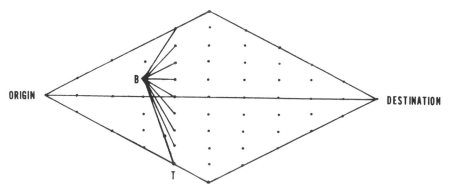

Figure 12.9.   Grid used in computerized OTSR scheme. (From Nagle, 1972.)

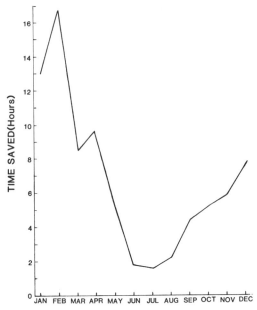

Figure 12.10.   Time saved over a great circle route by a 15-knot vessel sailing between Bishop Rock, England, and Norfolk, Virginia, based on 300 monthly voyages simulated using a computerized OTSR scheme. (After Nagle, 1972.)

that produces the optimal track across the whole ocean. The computer continually updates its sea state information by automatically reading the latest wave forecasts into its memory. The calculation itself follows the method used in the manual technique, and has the benefits of being less subjective, more rapid, and able to handle a number of ships at one time. Computerized OTSR has been found to be especially suited to a large military or commercial shiprouting operation. The technique is also ideal for repeated simulations used to determine the time saved by computerized OTSR on a particular voyage route. The results of such a simulation are shown in Figure 12.10, where the potential time saved for various months of the year is given for a voyage between Bishop Rock, England, and Norfolk, Virginia. The simulation revealed that the greatest savings are achieved for westbound voyages in the winter, and the least savings are for eastbound voyages in summer. Also, it shows that larger, faster ships are less adversely affected by high seas and therefore offer less of an opportunity for saving time by the use of OTSR. The use of more advanced computers, coupled with real-time meteorological and oceanographic observations from satellites, improved and more efficient computer wave models, and a better understanding of ship behavior in various sea conditions, will make computerized OTSR even more useful to marine transportation in future years.

12.4. DANGEROUS WAVES IN WESTERN BOUNDARY CURRENTS

In strong, persistent western boundary currents such as the Gulf Stream, warm ocean water often comes into contact with cold air masses. When this occurs two processes combine to produce a dangerous sea state that is a major hazard to marine navigation. The first is the interaction of the strong current with ocean waves being produced by winds blowing in the opposite direction to the current. This causes a substantial increase in wave height. The second is the interaction of the cold polar air with the warm ocean waters, which produces an increase in atmospheric instability and a subsequent increase in wind speed.

Figure 12.11 illustrates the meteorological conditions that typically exist when the joint effects of opposing currents and atmospheric instability are acting to increase wave heights. In this example, cold northerly winds were blowing into the warm Gulf Stream. The air, which was slightly unstable over the Slope water, became extremely unstable over the Gulf Stream, where a 10° to 12°C (18° to > 20°F) temperature difference existed between sea and air. The instability caused mixing of the high-speed upper air winds and the surface winds. The surface wind speed therefore increased, in this example, from ~ 60 knots (~ 30 m/s) (75 percent of the geostrophic value) in Slope water region to up to ~ 75 knots (~ 38 m/s) (90 percent of the geostrophic value) near the North Wall (see Table 3.1). The higher winds increased the significant wave height from ~ 25 ft (~ 7.6 m) in the Slope water to 30 ft (~ 9.2 m) near the North Wall. In addition to the instability effect, the opposing current also increases wave heights. James (1974) calculated that for an opposing current of 3 to 5 knots (1.5 to 2.5 m/s) the wave height would increase by 20 to 40 percent. The combined effects of atmospheric instability and opposing current in this case produced significant wave heights of 35 to 40 ft (10.7 to 12.2 m) along the North Wall where current speeds reach 3 to 5 knots. The steep and rapidly growing waves that occur along the North Wall of the Gulf Stream represent an extreme hazard to navigation. A number of ships have foundered off the coast in this and other western boundary currents under similar conditions. Dangerous waves along western boundary currents are therefore an important consideration for OTSR in these regions.

12.5. THE USE OF CURRENTS TO IMPROVE SHIP EFFICIENCY

Benjamin Franklin was among the first to make mariners aware of the benefits of favorable ocean currents when he printed a chart of the Gulf Stream in 1769. Over 200 years later nautical publications such as the *Coast Pilot* recommend shipping routes based on the average positions of major current systems, as illustrated in Figure 12.12. Although the average position of the Gulf Stream is well known, large variations in its location and strength are

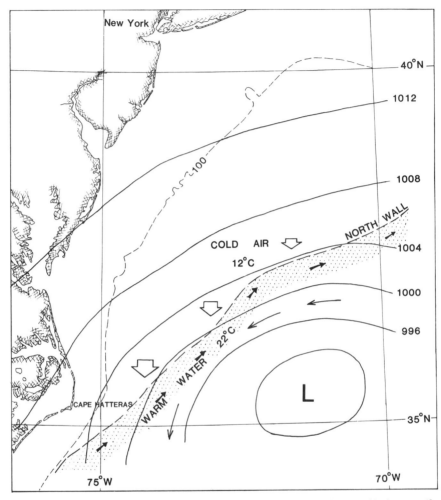

Figure 12.11. Marine weather conditions on March 27, 1971, showing a cold air mass (fat arrows) blowing over the warm (≥ 22°C) northeastward-flowing Gulf Stream water (dark arrows) and the opposing wind (thin arrows) circulating around the low pressure system. Atmospheric pressure is in millibars and sea surface temperature is in °C. (After James, 1974.)

observed over time periods on the order of days, as shown in Figure 12.13. It is apparent that the average position of the Gulf Stream should be considered only as an approximate guide to navigation and that real-time charts of this current would be more useful to the mariner because of the current's inherent variability. To meet this need oceanographers working with the shipping industry have produced synoptic charts of the Gulf Stream for marine navigational purposes.

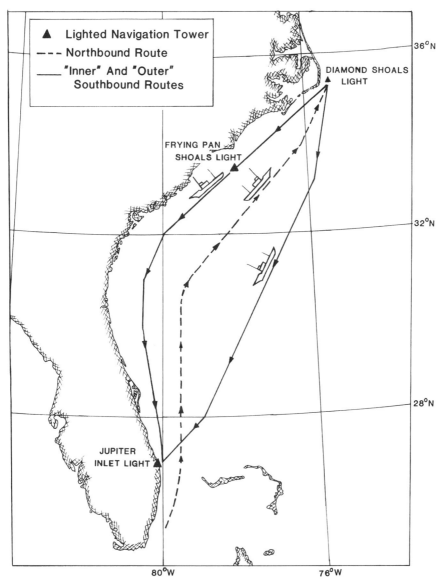

Figure 12.12. Vessel routes near the Gulf Stream recommended by the *Coast Pilot*. (After Bishop, 1978.)

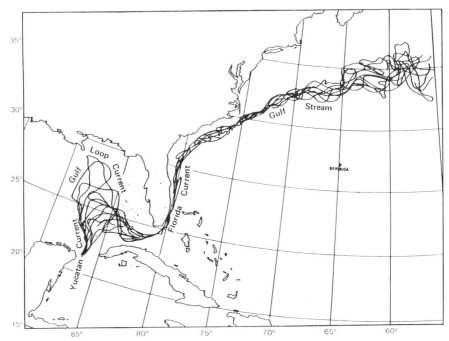

Figure 12.13. Variability of the position of the Gulf Stream current system. (After Maul, 1976.)

To determine the advantages of using this oceanographic analysis to improve ship efficiency, the EXXON Corporation and the National Oceanic and Atmospheric Administration (NOAA) carried out a cooperative fuel savings experiment. The program, initiated in 1975, tested whether the real-time Gulf Stream analysis leads to significant fuel savings. NOAA oceanographers studied Gulf Stream cross sections constructed from subsurface temperature data and located the current's core by finding the 15°C isotherm at a depth of 200 m. Satellite images of sea surface temperature were also analyzed. The strong surface temperature gradient representative of the North Wall was identified and plotted. Combining this information with available ship reports, the oceanographers produced synoptic charts of the Gulf Stream such as that shown in Figure 1.11. EXXON arranged for radio broadcast of this information to their tankers at sea. Eleven EXXON vessels sailing between the Gulf of Mexico and the east coast of the United States participated in the experiment. Five were instructed to follow the *Coast Pilot* route, using the average location of the Gulf Stream northbound and avoiding it southbound. The other six used the real-time NOAA data. Satellite imagery, surface and subsurface temperature data, and ship reports were analyzed twice a week to locate the North Wall. Radio broadcasts of the longitude at which the North Wall crossed each degree of latitude from

TABLE 12.1.   FUEL SAVINGS DURING THE NOAA/EXXON EXPERIMENT

Vessel	Average Speed	Barrels/Hour	Trips	Barrels Saved[a]
Baton Rouge	16.7	27.0	23	2,140
New Orleans	16.2	26.8	26	2,475
Boston	15.2	23.3	23	2,029
Lexington	17.5	29.6	22	2,142
Jamestown	17.7	28.3	26	2,392
Huntington	14.3	16.7	20	1,344
				12,522 Total[b]

Additional savings were calculated as follows:

Time	Additional Vessels	Trips	Barrels Saved
Sep.–Dec. 1975	5	121	10,619
Early 1976	4	97	8,382
Projected			
All 1976	15	362	31,523[c]

Source: J. Bishop (1978).

[a]All estimates based on average round-trip saving of 57.55 nautical miles.

[b]Total barrels of fuel saved by 6 ships in 140 transits during 7 months (Feb.–Aug., 1975) using NOAA satellite data.

[c]@$11.50/barrel, $360,000.

27°N to 38°N were sent to the six EXXON tankers. The ships' officers plotted these locations and shaped their courses accordingly. A summary of 140 transits using the NOAA data and 82 without it indicated an average round-trip savings of ~ 58 nautical miles. Table 12.1 shows the actual and projected fuel savings for the six tankers during the experiment. The EXXON projection for the full year, based on a total of 15 vessels using the analysis, was over 31,000 barrels of oil saved, which, using the 1976 price of oil, would amount to a potential annual savings of over 350,000 dollars. The use of satellite-derived Gulf Stream positions has become a standard navigational procedure for all ships of the EXXON Corporation. A similar practice for all coastal vessels navigating in these waters would result in significant fuel conservation.

This procedure could even have been carried a step further by including the ships' officers more actively in the oceanographic analysis. For example, radiofacsimile equipment, which can receive satellite images and Gulf Stream analysis charts directly on board the ship, could be used in locating the Gulf Stream's North Wall. This information, combined with subsurface temperature data obtained from a shipboard expendable bathythermograph (XBT), would allow mariners with very little training to locate the Gulf Stream and shape their course accordingly. The application of a similar analysis to other western boundary currents, such as Agulhas Current near the Cape of Good Hope, would be another logical extension of this fuel savings experiment, especially for the supertankers that sail these waters between the Middle East and Europe.

Future advances in at-sea data collection and reporting systems, improved models of oceanic processes, and advances in remote sensing will allow oceanographers to better determine winds, waves, and currents over the world's oceans. With the application of this information, marine transportation will become safer and more efficient. As the economics of OTSR are realized by more of the worldwide shipping industry, more advanced techniques for routing vessels will be developed that require even more detailed input information from physical oceanographers.

Thirteen

SEARCH AND RESCUE AT SEA

Maritime law and tradition require a master of a vessel to assist those in distress at sea. The search and rescue of those lost and in distress at sea represents a unique application of marine meteorology and physical oceanography to marine transportation. In this chapter the two aspects of search and rescue (SAR) that will be addressed are the relationship between the physical environment and SAR operations at sea and the techniques used to plan and coordinate these operations. Because the U.S. Coast Guard is a world leader in the advancement of SAR techniques, much of the discussion is based on their manuals and procedures.

The Coast Guard's involvement with search and rescue began in 1831, when the Secretary of the Treasury ordered the *Gallatin* to cruise the coast and provide assistance to mariners in distress. The Coast Guard now receives about 70,000 SAR calls a year, mostly during the recreational boating season. The searches themselves are planned using experience, an understanding of the marine environment, and statistical techniques. In addition to its involvement in coastal SAR operations, the Coast Guard also maintains the Automated Mutual Assistance Vessel Rescue System (AMVER), which is an international distress response system used for open ocean emergencies. This computer-based system annually tracks over 100,000 voyages of more than 6000 ships flying the flags of 75 countries, recording reported vessel positions and estimating the progress of each ship between reports. When a need for assistance arises, the AMVER controller obtains a print-out that provides a list of ships in the incident area, how they may be contacted, their SAR capabilities, and whether a doctor is aboard. The appropriate ship is then contacted and asked to assist the distressed craft.

ocean values of sea current, additional information is used by the Coast Guard to obtain estimates of sea current near the coast. One such technique is the weekly sea current chart shown in Figure 13.2. This chart is derived from an analysis of remote sensing data and is based on the relationship between sea surface temperature patterns and ocean current boundaries. The coastal current model published by Bishop and Overland (1977) and described in Chapter 2 was originally developed by the Coast Guard for coastal sea current estimates because the values given in atlas presentations were not considered reliable in shallow water. Sea current values are used in SAR planning for any incident of greater than 6 hours' duration, unless the target is in a high-speed current such as the Gulf Stream. In that case the sea current values are always used, regardless of incident duration.

Wind current.   This is the current generated by local winds acting on the ocean surface over a short period of time. Wind current is generally added vectorially to sea current to estimate the total surface trajectory of a distressed craft. In SAR planning wind current is considered to be a result of the present winds. The actual trajectory calculation generally uses wind forecasts or observations commencing 48 hours before the time of the incident and divided into 6-hour time periods. The method that the Coast Guard uses to calculate the wind current is based on a time-dependent form of the Ekman equations of motion. This model determines the drift contribution of the wind during each of the eight 6-hour time periods and then sums these vectorially over time. Figure 13.3 illustrates how this transient Ekman model is used to calculate the wind-drift current for SAR planning. The James curves for calculaing wind-driven currents (see Chapter 2) have also been used to calculate wind current for SAR purposes.

Leeway.   This is the movement of a search object resulting from the action of wind pushing against the exposed surface of the craft. The more surface area over which the wind blows, the greater the effect of leeway on drift. The craft's heading also influences leeway. A boat heading directly downwind or into the wind will have a smaller leeway than one heading crosswind. Leeway estimates based on empirical data collected by the Coast Guard are shown in Figure 13.4.

In calculating the datum for SAR planning purposes, surface drift is estimated based on a vector sum of sea current, wind current, and leeway, as shown in Figure 13.5(*a*). A problem arises with this approach in ocean regions with persistent winds because the sea current obtained from a current atlas already includes the effect of the constant local wind. Adding a local wind current to the wind-generated sea current might cause one to "double count" the effect of winds on the surface current. It is therefore common practice to omit the wind current from the datum calculation for such regions as the persistent trade wind belts. This is shown in Figure 13.5(*b*). In addition to calculated values, on-scene observations of currents can be used in

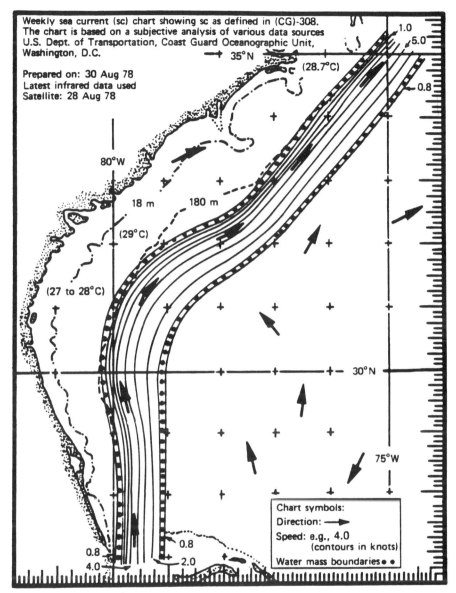

Figure 13.2. Weekly sea current chart used for SAR planning by the Coast Guard. (Courtesy of U.S. Coast Guard.)

211

WIND CURRENT COMPUTATION SHEET

VALID FOR TIME _151200Z-151800Z_   LAT. _44°-15'N_   LONG. _58°-25'W_

PERIOD	WIND HISTORY	COEFFICIENTS	CONTRIBUTION TO WIND CURRENT	LOCAL WIND CURRENT
1	260°T 35	221° 0.023	121°T 0.80	135°T 0.61
2	240°T 30	007° 0.010	247°T 0.30	
3	230°T 25	136° 0.007	006°T 0.18	
4	230°T 20	264° 0.006	134°T 0.12	
5	230°T 20	031° 0.005	261°T 0.10	
6	230°T 20	159° 0.004	029°T 0.08	
7	220°T 15	286° 0.004	146°T 0.06	
8	220°T 15	053° 0.004	273°T 0.06	

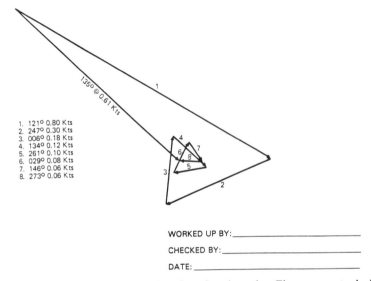

1. 121° 0.80 Kts
2. 247° 0.30 Kts
3. 006° 0.18 Kts
4. 134° 0.12 Kts
5. 261° 0.10 Kts
6. 029° 0.08 Kts
7. 146° 0.06 Kts
8. 273° 0.06 Kts

WORKED UP BY:_____

CHECKED BY:_____

DATE:_____

Figure 13.3.   Wind current computation sheet for a time-dependent Ekman current calculation as employed in SAR planning. The coefficients shown are the Ekman deflection angle and the appropriate Ekman wind factor. (From U.S. Coast Guard, 1973.)

SAR planning. A datum calculation using observed currents is shown in Figure 13.5(c). In some nearshore regions tidal currents are considered in SAR datum calculations. Because of their reversing nature, tidal currents are not generally used in long-term drift calculations, although they can become important in special cases, such as when target intercept time occurs at the time of maximum flood or ebb current. Longshore currents, caused by large incoming swell striking the shoreline at an angle, have also been considered in SAR planning for search objects within a mile or so from shore.

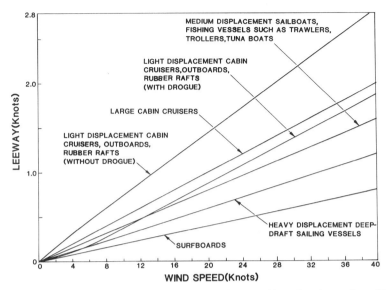

Figure 13.4. The relationship between leeway and wind speed based on data collected by the Coast Guard for various types of search targets. (From U.S. Coast Guard, 1973.)

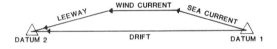

a. TYPICAL DATUM CALCULATION

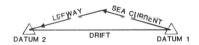

b. DATUM CALCULATION IN REGION
OF PERSISTENT WINDS

c. DATUM CALCULATION IN REGION
WHERE OBSERVED CURRENTS
ARE AVAILABLE

Figure 13.5. Three typical datum calculations. (From U.S. Coast Guard, 1973.)

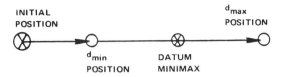

a. BASIC PLOT

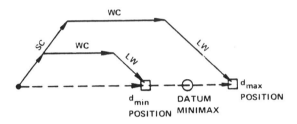

b. TIME UNCERTAIN

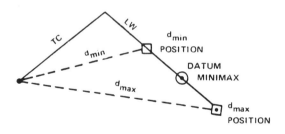

c. DROGUE/NO DROGUE   UNCERTAIN

Figure 13.6. Examples of minimax calculations, showing the maximum drift distance d_{max}, the minimum drift distance d_{min}, and the minimax datum, where SC = Sea Current, WC = Wind Current, TC = True Course, and LW = Leeway. (From U.S. Coast Guard, 1973.)

In many SAR incidents datum calculations are complicated by the lack of exact information on surface drift factors. The technique employed to address this problem is called minimax. It uses the minimum and maximum estimates for each drift factor to calculate a minimum and maximum datum point. The datum point used as a basis for the search operation is then established midway between these two positions. The basic minimax procedure is illustrated in Figure 13.6(a). The minimax approach is used in a number of situations, such as when there is doubt over the time that a lost craft has been adrift without power. This particular type of minimax calculation is illustrated in Figure 13.6(*b*). Also, if it is not known whether a drifting craft has a drogue to slow its drift, a minimax calculation of the type shown in Figure 13.6(*c*) is used.

Once the datum location is established, the error in the datum position must be determined. The size of the error in the datum location determines the size of the planned search area. In general, the larger the search error, the greater the area that needs to be searched. In SAR planning the three major components of the search error E are the total drift error De, the initial position error X, and the search craft error Y. The formula used to compute the total probable error is

$$E = \sqrt{De^2 + X^2 + Y^2}$$

The total search error includes the accumulated drift errors over the total SAR mission. By experience search planners have found that a factor of one-eighth of the total drift is a good approximation for the drift error. The second component of the search error, the error in the initial position of the object, is derived from an estimate of the navigational accuracy of the distressed craft. The third component of the search error, error introduced by the search craft, is calculated based on the navigational accuracy of the search craft.

When a minimax datum calculation is used a graphical or an algebraic method can be employed to find the drift error. In the graphical method the error in a particular drift calculation de is found for both the minimum drift (d_{min}) and the maximum drift (d_{max}) by dividing each by eight. Using the positions of d_{min} and d_{max} as the center, two circles are drawn, one with a radius of $d_{min}/8$ and the other with a radius of $d_{max}/8$. A third circle, as illustrated in Figure 13.7, is drawn with a radius $de_{minimax}$ representing the minimax drift error for a particular drift calculation. The total drift error De is the sum of each calculated drift error de during the total SAR mission. In the algebraic method the minimum and maximum positions are plotted on a navigational chart and the distance D between the two positions is measured (see Figure 13.7). The drift error is then calculated as

$$de_{minimax} = \frac{D + de_{min} + de_{max}}{2}$$

$$de_{min} = d_{min}/8$$

$$de_{max} = d_{max}/8$$

As an example of this calculation consider a small boat which failed to return from a day of fishing. The owner was known to have been planning to fish in the vicinity of a particular oil rig and was last seen 8 hours before he was due home. It was determined that the boat probably broke down sometime in

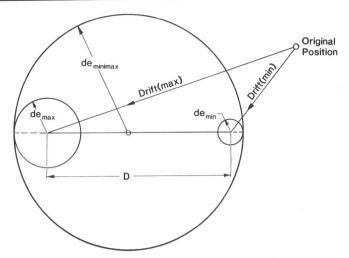

Figure 13.7. Graphical technique used to calculate the minimax drift error, $de_{minimax}$. Shown are the drift error for the maximum drift calculation de_{max}, the drift error for the minimum drift calculation de_{min}, and the distance D between the maximum and minimum drift positions. (From U.S. Coast Guard, 1973.)

that 8-hour period. Using the latest time of probable breakdown the minimum drift is calculated as 30 km, while the maximum drift, calculated from the earliest time of probable breakdown, is 86 km. The minimax drift error about the datum point would therefore be

$$de_{minimax} = \frac{56 + \dfrac{30}{8} + \dfrac{86}{8}}{2} = 35.3 \text{ km}$$

13.2.2. Defining the Search Area

After the datum location has been established and the total search error determined, the search area must be defined. The area within a circle with its center at datum and with a radius equal to the total probable error E has a 50 percent probability of containing the search target. To ensure a greater than 50 percent probability that the target is in the search area the radius of the search area is taken as the total probable error plus an additional length. The radius of this circle is called the search radius R and is computed as the product of the total probable error and a search safety factor f_s:

$$R = E \cdot f_s$$

where $f_s = 1.1$ for the first search, 1.6 for the second search, 2.0 for the third, 2.3 for the fourth, and 2.5 for the fifth search. After the search radius, including the safety factor, is computed, the search area can be defined, as

Search Areas—Stationary Datum Point

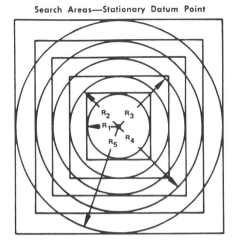

Search Areas—Moving Datum Point

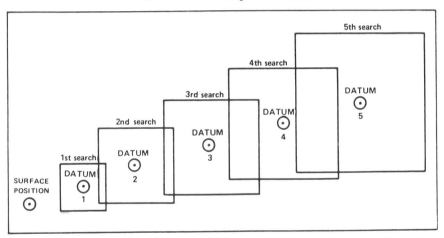

Figure 13.8. Calculation of search areas for a stationary and a moving datum. (From U.S. Coast Guard, 1973.)

shown in Figure 13.8 for a stationary datum and for a moving datum. Note that the search areas are rectangular, not circular, since this is a more practical search pattern for actual search operations. Since the total probable error and the safety factor increase from search to search, the search area increases in size with time.

13.2.3. Selection of a Search Pattern

After the location and size of the search area have been determined an appropriate search pattern is selected based on the accuracy of the datum, size of the search area, navigational accuracy, weather and sea conditions, size of the target, and the types of detection aids the survivors may have. The type and number of available search units will also influence the selection of search pattern. Search patterns are chosen with the intention of locating survivors as rapidly as possible by using a search-track spacing based on radio, visual, or other signaling aids that the survivors possess. Searchers must keep in mind that the survivors will be in better physical condition to use these aids during the initial periods of the search and that the battery life of locater beacons is limited to between 24 and 48 hours. Three commonly used SAR search patterns are as follows.

Trackline search.   This pattern is used when a vessel or aircraft is missing somewhere along its intended trackline, the assumption being that if the distressed craft is near her intended route her crew will be able to signal the search craft. The trackline search pattern, illustrated in Figure 13.9, permits a rapid and thorough coverage of the missing craft's track and the area immediately adjacent to it.

Parallel track search.   This pattern is used when the search area is large, only the approximate location of the target is known, and uniform search coverage is desired. A single-unit parallel track search pattern, as illustrated in Figure 13.9, is best suited for rectangular search areas with search legs aligned parallel to the major axis of the search area. Parallel track patterns are used in searches in which navigational aids give adequate coverage of the search area.

Sector search.   This search pattern is used when the position of the search target is known with relative accuracy and the area to be searched is not extensive. A single-unit sector search pattern, as illustrated in Figure 13.9, resembles the spokes of a wheel and covers a circular search area. A datum marker such as a smoke marker, a radio beacon, or a radar beacon, can be used to mark the center of the search area and as a navigation aid on each leg of the search. Compared with the trackline or parallel search patterns, the sector search is not only easier to execute and navigate, it is also more effective. The track spacing, which is small near the center, ensures increasingly intensive coverage in the region where the target is most likely to be found.

13.2.4.   Detection of a Target

Detection of a search target depends on the track spacing S, the probability of detection P, the sweep width W, and the coverage factor C. Track spacing is the distance between adjacent search tracks, which may be simultaneous sweeps by several search units or successive sweeps by a single unit. Closer track spacing will lead to a greater chance of spotting the target but will cause an increase in search time for a fixed number of search units, so that a

Trackline Pattern

Parallel Track Single-Unit

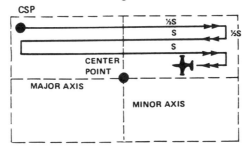

Sector Single-Unit

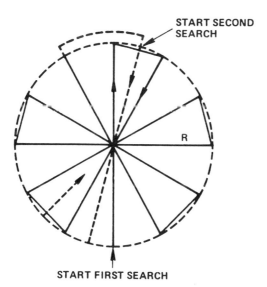

Figure 13.9. Three commonly used oceanic search patterns. CSP is the Commence Search Point and S is the track spacing. (From U.S. Coast Guard, 1973.)

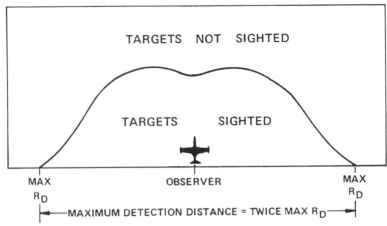

Figure 13.10.   The relationship between the instantaneous probability of detection (IP) and the sighting of targets for a visual search by aircraft. (From U.S. Coast Guard, 1973.)

trade-off must be made between track spacing and the size of the search area to be covered. The optimal track spacing is that which permits the maximum expectation of target detection in the available time consistent with the economical deployment of available search units.

Instantaneous probability of detection (IP) is the ratio of targets detected to targets missed for each scan made by the search unit's lookout or detection equipment. The IP determines the probability of detection over successive scans as the search unit moves along its track. In general the IP is highest at short lateral distances from the search unit and decreases as the distance increases. Figure 13.10 illustrates the typical decrease of the IP with distance for a visual search from an aircraft.

Sweep width, illustrated in Figure 13.11, is a measure of detection capability based on target characteristics, weather conditions, and limitations of detection equipment. It is the numerical value obtained by reducing the maximum detection distance of a scan because of on-scene conditions. Sweep width is defined as the area within which the number of targets missed equals the number of targets sighted, and is always less than the maximum possible detection distance. Tables of sweep width values for

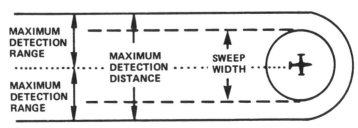

Figure 13.11.   Sweep width and its relationship to the maximum detection distance. (From U.S. Coast Guard, 1973.)

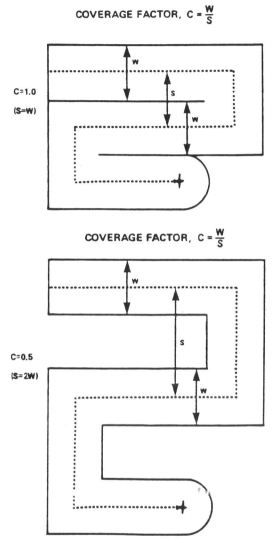

COVERAGE FACTOR, $C = \frac{W}{S}$

C=1.0

(S=W)

COVERAGE FACTOR, $C = \frac{W}{S}$

C=0.5

(S=2W)

Figure 13.12. Two search patterns, one with coverage factor $C = 1.0$ and one with $C = 0.5$. (From U.S. Coast Guard, 1973.)

various environmental conditions and types of searches, including visual, radar, and electronic, have been derived from data collected by the Coast Guard.

Coverage factor is a measure of the search effectiveness and is defined as $C = W/S$. Figure 13.12 illustrates two searches, one with a coverage factor C of 1.0 and a second with a coverage factor C of 0.5. A higher coverage factor indicates a more thorough coverage of the search area. The coverage factor therefore determines the probability of detection for a given search and for repeated searches. The relationship between coverage factor and probability

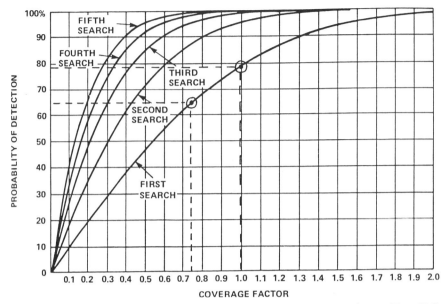

Figure 13.13. Relationship between probability of detection and coverage factor. (From U.S. Coast Guard, 1973.)

of detection is shown in Figure 13.13. As an example, consider a search coordinator who plans for a probability of detection $P = 78$ percent during the first search with a sweep width $W = 3$ km. Using Figure 13.13, with $C = 1.0$, leads the coordinator to choose a track spacing $S = W/C = 3$ km to obtain the 78-percent probability of detection. If the search unit is unable to use this track spacing and instead uses $S = 4$ km, the actual coverage factor $C = W/S$ is 0.75, which changes the probability of detection to 65 percent.

It is the responsibility of the search coordinator to combine effectively the elements of a search plan described in this chapter. A key factor in doing this is understanding the operational environment. Knowledge of marine weather conditions, ocean currents, sea state, and even sea water temperatures forms the basis for decisions that may determine the ultimate success of a SAR operation. In the future SAR planning, based on the documented success of SARP and CASP, will become more automated and will use input data derived from satellite-relayed distress signals. Also, improvements in marine weather forecasts and new improved models of wind waves and ocean currents could have increased application in SAR planning. To realize the advantages of these improvements physical oceanographers and marine meteorologists need to work more closely with SAR planning personnel to tailor their efforts to the needs of the SAR coordinator. When this occurs the full benefits of these scientific advances will more closely fit user needs in this important area of marine transportation.

Fourteen

MILITARY USES OF THE OCEAN

Meteorological and oceanographic information are important for a number of tactical and strategic military uses. Although all aspects of the physical environment, including weather conditions, waves, tides, and currents, are important for various operational applications, recent emphasis has been on the effects of the physical environment on SONAR (Sound Navigation And Ranging) equipment used for submarine detection. Submarine-carried nuclear missiles are an important part of the defense systems of the superpowers and may even be considered to be a stabilizing factor in the global balance of nuclear power. Because of the inherent difficulties in locating submarines, missiles which can be launched from these platforms act as a deterrent to nuclear attack on the nation itself.

Military operations intended to detect and attack submarines are known as anti-submarine warfare (ASW). SONAR, used in either an active or a passive mode, is the primary tool for ASW operations. Active systems transmit underwater sound pulses that are reflected by a target and then received back at the transmitting site, whereas passive systems merely receive sounds made by a target. Active SONAR systems can determine the range and bearing to a target, whereas a single passive system can only determine the bearing. On the other hand, passive systems do not make detectable sounds and are effective over longer ranges than are active systems.

Much progress has been made in the development of ASW systems since World War II, although the most efficient use of modern SONAR equipment still requires a synoptic picture of the oceanic thermal structure. Until the oceanic state can be mapped in near real time, submarines will continue to move for the most part undetected by SONAR through large portions of ocean space. In this chapter the relationship between physical oceanography and selected military uses of the ocean will be given. A general discussion of techniques used by naval forces to predict the oceanic thermal structure, which is a key element in determining SONAR performance, will be included. In addition, uses of ocean wave information for amphibious operations and aircraft ditching procedures will be outlined, as will be the relationship between submarine diving operations and the vertical temperature structure.

14.1. UNDERWATER ACOUSTICS AND ASW OPERATIONS

A major area of interest to naval oceanographers is underwater acoustics, the study of which provides information that aids in the use of SONAR. Since the velocity of sound in the ocean depends on water temperature, salinity, and pressure, SONAR can be used most effectively only when the distributions of these parameters, especially that of temperature, are known. In the upper ocean, for a temperature of 0°C and a salinity of 35 ppt, the speed of sound is ~1450 m/s. This speed increases by ~1.3 m/s for every 1-ppt increase in salinity, by ~4.5 m/s for every 1°C increase in temperature, and by ~1.7 m/s for every 100-m increase in depth, because of pressure effects.

Temporal and spatial variations in temperature, both horizontal and vertical, especially those occurring near ocean fronts or the thermocline, greatly affect the velocity and propagation of sound, as shown in Figure 14.1. In the region of the thermocline the temperature effect dominates and sound velocity usually decreases rapidly with depth. Below approximately 1000 m the water becomes isothermal and the pressure effect is dominant. Here sound velocity increases with increasing depth, and because of refraction, horizontally propagated sound waves will be bent downwards in the thermocline region and upwards in the deeper regions. This causes sound waves to be trapped and concentrated in a region approximately 1000 to 1500 m deep, where sound velocity is at a minimum. This region is called the sound channel, since it provides an effective medium for the horizontal propagation of sound waves over very long distances. This effect is important for SOFAR (Sound Fixing And Ranging) applications (Fig. 14.1, top left). Other sound transmission modes, related to vertical variations in temperature and resulting vertical sound velocity profiles, are also shown in Figure 14.1. The two most important sound transmission effects for naval applications are the sound channel, in which distress signals can be transmitted for thousands of kilometers, and shadow zones, in which sound waves bend radically. Submarines within a shadow zone can operate undetected by SONAR. Sound transmission can be predetermined by using observations and forecasts of

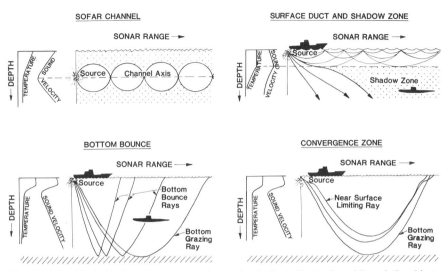

Figure 14.1. Variation in temperature and sound velocity with depth and its relationship to sound transmission. (After U.S. Naval Oceanographic Office, 1965.)

the oceanic thermal structure. This information is therefore critical in making tactical ASW plans at sea.

14.2. MANUAL OCEAN THERMAL STRUCTURE FORECASTING

Physical oceanographers have developed both manual and computerized techniques to predict ocean thermal structure for ASW applications. Initially, ocean thermal structure forecasting was largely subjective, using on-scene manual calculations based on empirically derived graphs and equations. In recent years these techniques have become more objective, using relationships based on the equations of motion and computers to make the required calculations. The manual procedure for ocean thermal structure forecasting is based on empirical relationships among air–sea interaction processes and uses on-scene input data to provide forecasts in an ASW operating area for up to 48 hours. Although this method depends on empirical relationships and at times suffers from a lack of accurate input data, it has been used effectively under operational conditions to calculate short-term changes in thermal structure for ASW applications.

The oceans are, as is described in Chapter 1, characterized by a surface mixed layer, a thermocline region, and a deep layer. Depending on atmospheric conditions, the temperature in the surface layer can be isothermal, can decrease, or can even increase slightly with depth. Clear skies and light winds may result in the formation of a strong thermal gradient near the sea surface, called the afternoon effect, which causes diurnal variations in oceanic sound propagation and therefore is important to ASW operations. In addition, a number of oceanic processes that involve the transfer of heat and momentum, including insolation, which is direct heating of the sea by solar radiation; back radiation from the ocean, which occurs mainly at night; evaporation, which decreases sea surface temperatures; surface heat flux; convective and turbulent mixing; and advection affect the thermal structure of the ocean. Internal waves, described in Chapter 4, complicate the observation and prediction of the upper ocean thermal structure because the thermocline rises and falls with the passing of these waves. ASW thermal structure forecasts must be used with caution in areas where internal waves are believed to be present.

The suggested steps given by James (1966a) for making an on-scene manual ocean thermal structure forecast are illustrated in Figure 14.2. These steps, each of which is based on empirically derived relationships for heat and momentum transfer processes in the oceanic surface layers, are as follows.

1. *Establish Existing Conditions.*   To make a prediction of the future ocean thermal structure requires an estimate of present conditions. Recent observations, extrapolation from previous observations, nearby observations, or data from climatological atlases are used to define the existing oceanic thermal structure.

2. *Compute Advection.*   Advection by local currents may have an important influence on the ocean thermal structure. The size of this effect will depend on the length of the forecast period, the permanent circulation pattern, and the wind conditions. If the permanent circulation is in geostrophic balance, the flow pattern will tend not to modify the existing thermal structure. Wind-drift currents, on the other hand, may have a strong influence on the ocean thermal structure, especially when flowing perpendicular to surface isotherms. Wind-drift advection is calculated using the James wind-drift current curves described in Chapter 2. Empirically derived relationships between advection and changes in thermal structure are then used to compute this component of the forecast.

3. *Compute Heat Budget.*   The influence of heat exchange on ocean thermal structure is computed using empirical equations involving solar elevation, insolation, back radiation from the sea surface, evaporation and condensation, and the surface heat flux. James (1966a) calculated that these processes may cause as much as 1 to 2°C (1.8 to 3.6°F) change in the surface water temperature, although the effect decays rapidly within the first 20 m of the water column.

4. *Compute Mixing.*   A net heat gain at the sea surface produces a warming of the surface waters. When this occurs it is necessary to calculate the depth to which the turbulent action of the wind will mix these warmed surface waters. A net surface heat loss, on the other hand, results in a

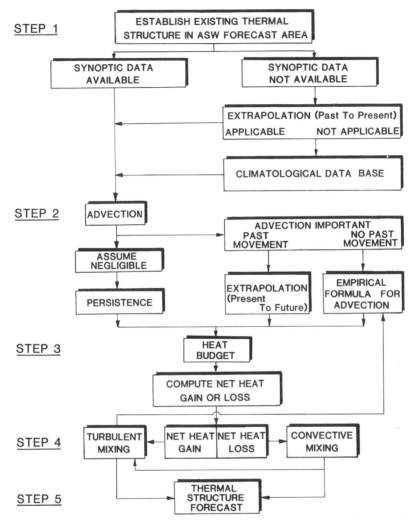

Figure 14.2. The steps in making a James manual ocean thermal structure forecast. (After James, 1966a.)

decrease in the surface temperature and convective mixing of these denser surface waters. Under this condition both convection and turbulent mixing must be considered. Figure 14.3 illustrates various surface processes and their relationships to the ocean thermal structure. In the manual method, the magnitude of each process is calculated from empirically derived graphs and equations.

   5. *Presentation of Results.* The final step in the manual procedure is to use the results of the initial steps in a forecast, which is then categorized into one of three time periods. The first is the time period for which persistence

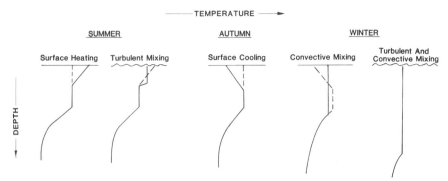

Figure 14.3. Variations in surface thermal structure due to surface heating and cooling and turbulent or convective mixing. The dashed lines represent the temperature profile before and the solid line the profile after each process. (After James, 1966a.)

from available observations is considered the best forecast; the second is the time period for which the manual calculation of heat budget, mixing, and advection should be used; and the third is the time period for which climatology provides the best forecast. The duration of each period is determined by the forecaster and depends on many factors, including the previous validity of the empirical equations used in the manual calculations and the experience of the forecaster. Persistence is usually considered the best forecast for the first day or so after a set of observations has been taken. This consideration depends on the estimated magnitude of the local variations caused by heat exchange, mixing, and advection. In general, when the expected variations in the thermal structure exceed the inherent error in the observations, a manual calculation is made. After the computed forecasts lose reliability, say after a few days to a week or so, climatology becomes the best forecast of the thermal state. These time periods vary in length among operating areas, so the forecaster must determine for how long persistence gives the best forecast, when to use computed changes, and at which time climatology becomes valid.

Each component of the manual method is based on empirical relationships and requires reliable input data for accurate forecasts. The method is largely subjective, and although it has proved to be a useful tool, the time required to make the calculations limits its usefulness. The manual ocean thermal prediction technique uses climatological data combined with empirical equations and real-time observations. Its success has been limited primarily because of the large variability in the ocean thermal structure about its climatological mean. Also, traditional oceanic observation systems are not capable of resolving oceanic variability on the temporal and spatial scales sometimes required for ASW applications. The use of remote sensing data from both aircraft and satellites in combination with conventional observation methods will improve this capability in the future.

14.3. COMPUTERIZED OCEAN THERMAL STRUCTURE FORECASTING

Computerized ocean thermal structure forecasting is based on the equations of motion in combination with sophisticated data analysis techniques. The United States Navy has developed a model called the Thermodynamical Ocean Prediction System (TOPS) which is operated by the Naval Fleet Numerical Oceanography Center at Monterey, California. The TOPS model is classified as a synoptic mixed-layer dynamical model that includes oceanic thermodynamical processes and, as reported by Clancy et al. (1981), uses oceanographic data from the Navy's worldwide data base. The predicted surface wind field is the most important input to the TOPS model. This parameter is routinely calculated by the Navy's global weather forecasting model. TOPS model outputs are produced on a 63×63-point grid for the Northern Hemisphere, as shown in Figure 14.4. This figure also shows the vertical grid for TOPS, which is defined such that temperature, salinity, and currents are calculated at 17 levels from the ocean surface to a depth of 500 m. The vertical grid allows the calculation of small-scale surface features, such as the afternoon effect, that are important to ASW operations. Figure 14.5 illustrates the afternoon effect, showing time–depth contours of sound speed for typical diurnal variations of the surface thermal structure. The TOPS model produces a 3-day forecast of the upper ocean thermal structure which is used directly for tactical ASW operations by naval forces at sea.

The Navy has tested TOPS a number of times under operational conditions. These tests were conducted in the North Atlantic and North Pacific Oceans, regions characterized by strong storm systems and where adequate observational data were available for model verification. Statistics of model forecast errors, defined as modeled temperature values minus observed values, were computed as a measure of the model's performance. These were compared to the error in a forecast based on persistence alone to determine if the model represented a significant improvement. One example of a typical synoptic condition under which this model comparison was done is shown in Figure 14.6. The region of model calculation, shown as a rectangular area on the synoptic chart, represents a 3- $\times$ 5-point subset of the 63- $\times$ 63-point TOPS grid. This region was used in the test because it represented storm conditions and was relatively data rich. Figure 14.6 also shows, as an insert, the root-mean-square forecast error for persistence versus that for TOPS during this model comparison.

Modern ASW operational planning uses both on-scene manual and shore-based computerized techniques for making ocean thermal structure forecasts. Both approaches have advantages and disadvantages, and are considered compatible systems by the Navy. Dynamical methods such as TOPS provide a rapid, relatively accurate calculation based on the Navy's large data base, with calculations being done at a shore-based computer facility.

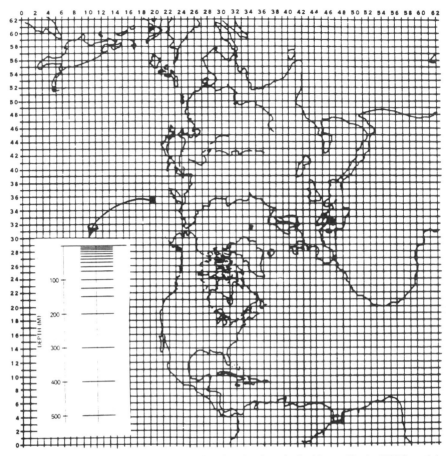

Figure 14.4. Standard 63 × 63 horizontal and 17-level vertical grids used in the TOPS model. (From Clancy et al., 1981.)

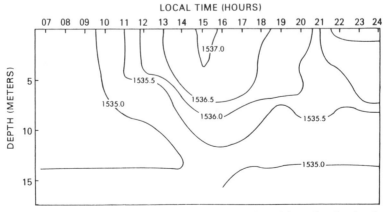

Figure 14.5. Variation of sound speed (m/s) with depth and local time, showing the afternoon effect. (From Shonting, 1964.)

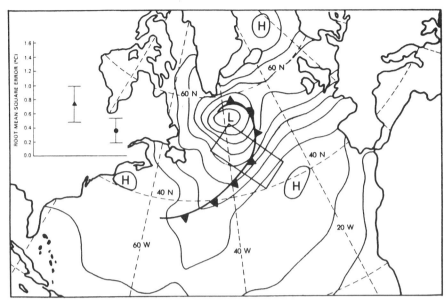

Figure 14.6. Surface meteorological conditions on August 30, 1980, during a test of the TOPS model. Also shown is the root-mean-square error (°C) for a TOPS forecast (●) and for a 3-day persistence (▲) forecast for the test. Error bars shown represent 95-percent confidence limits. (From Clancy et al., 1981.)

This could be a disadvantage in times of war, when model input data or model output transmissions to at-sea units may be greatly reduced. The manual method, which was developed for shipboard use, requires a knowledge of physical oceanography, adequate local oceanographic data, and time to make the analysis and forecast. It represents an important additional tool that on-scene forces can use for ASW operational planning even under wartime conditions. The manual and computerized ocean thermal structure forecasting methods therefore both represent parts of an effective ASW operations plan.

14.4. APPLICATIONS OF OCEAN WAVE INFORMATION

Knowledge of the sea state is important for a number of aspects of military operations at sea. Oceanographers, for example, helped in the planning of the major Allied amphibious operations during World War II by using the Sverdrup–Munk significant-wave technique to estimate wave and surf conditions. The use of the Sverdrup–Munk method marked one of the first applications of oceanography to modern military operations at sea. In addition to a surf forecast, other critical information such as tidal height and nearshore current forecasts was needed by the military commander for each

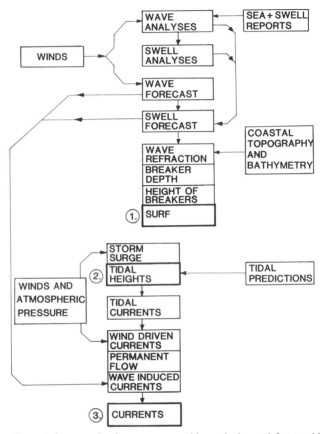

OCEANOGRAPHIC ANALYSIS

PROCEDURE FOR AMPHIBIOUS OPERATIONS

Figure 14.7. General framework of an oceanographic analysis used for amphibious operations, showing (1) surf prediction, (2) tidal height, and (3) currents as the three primary outputs needed by the military commander. (After U.S. Navy, 1965.)

amphibious operation. Figure 14.7 outlines the analysis procedure and information typically produced for a military commander during an amphibious operation. The final approval for the Allied invasion of North Africa and France was given only after an extensive analysis of wind, wave, tide, and current conditions was made.

The application of information on ocean wave conditions is also of vital importance for ditching operations of aircraft at sea. The great danger during ditching is the rapid deceleration as the aircraft hits the water, resulting in damage to both the plane and its crew. This danger can be minimized by selecting the best possible ditching heading based on wave conditions. When

TABLE 14.1. PORTION OF THE BEAUFORT WIND SCALE USED TO ESTIMATE WIND SPEED FOR DITCHING OPERATIONS

Beaufort Number	Wind Velocity (Knots)[a]	Sea State	Wave Height (Ft)[b]
0	1	Like a mirror	0
1	1–3	Ripples with the appearance of scales	$\frac{1}{2}$
2	4–6	Small wavelets; crests have a glassy appearance and do not break	1
3	7–10	Large wavelets; crests begin to break Foam of glassy appearance; few scattered whitecaps	2
4	11–16	Small waves, becoming longer; fairly frequent whitecaps	5
5	17–21	Moderate waves, taking a pronounced long form; many whitecaps	10
6	22–27	Large waves begin to form; white foam crests are more extensive; some spray	15
7	28–33	Sea heaps up and white foam from breaking waves begins to be blown in streaks along the direction of waves	20
8	34–40	Moderately high waves of greater length; edges of crests break into spindrift; foam blown in well-marked streaks in the direction of the wind	25
9	41–47	High waves; dense streaks of foam; sea begins to roll; spray affects visibility	30
10	48–55	Very high waves with overhanging crests; foam in great patches blown in dense white streaks along the direction of the wind; whole surface of sea takes on a white appearance; visibility is affected	35

Source. U.S. Coast Guard (1973).
[a]1 knot ≃ 51.4 cm/s.
[b]1 ft ≃ 30.5 cm.

ditching is being considered, the pilot analyzes the sea state from as high an altitude as visibility permits. The most dangerous swell can readily be distinguished at an altitude of ~1000 m. As the aircraft descends wind direction and speed are also determined from the sea state. Wind streaks on the sea surface usually line up with the wind direction. Whitecaps fall forward with the wind, but are overrun by the longer waves, producing the illusion that

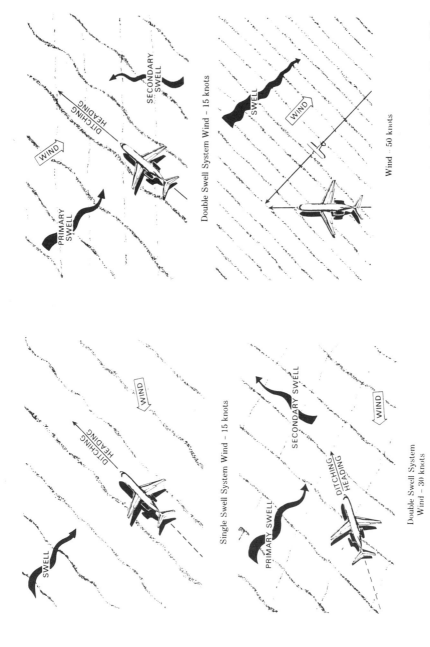

Figure 14.8. Combinations of wind (sea) and swell conditions that define various procedures for aircraft ditching. (From U.S. Coast Guard, 1973.)

234

the foam is sliding backward. This knowledge is applied by the pilot to determine the wind direction. Wind speed is estimated from the appearance of the whitecaps, foam, and wind streaks and using the Beaufort wind scale, which is given in Table 14.1. After the wind, sea, and swell conditions have been determined, the pilot chooses the proper heading. Some of the possible combinations of sea conditions and headings for ditching are shown in Figure 14.8. A determination of sea and swell conditions is critical to a successful ditching operation.

14.5. DENSITY STRATIFICATION AND SUBMARINE OPERATIONS

The normal operation of a submarine, which requires the vessel to move through layers of different density, can be accomplished by changing the amount of water ballast aboard the ship. However, in regions with strong density gradients, especially near the thermocline and oceanic fronts, surfacing and diving can become complicated. When diving through a strong thermocline a submarine will encounter water of increased density. This may produce enough buoyancy to counteract the normal loss of buoyancy resulting from pressure-induced hull compression. Figure 14.9 illustrates how three different vertical temperature profiles may affect submarine ballasting during a dive. Temperature profiles for an isoballast dive, one with no flooding or pumping of water, are also shown. The vertical temperature profile given by the dashed line in Figure 14.9 shows the commonly observed surface isothermal mixed layer over a sharp thermocline. Consider a submarine operating in these waters that starts a dive at periscope depth (loca-

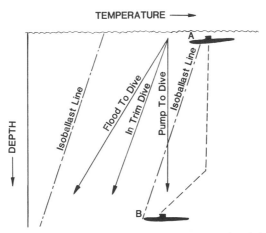

Figure 14.9. Vertical temperature profiles (solid lines with arrows) and their relationship to submarine diving operations. Isoballast lines show temperature profile that will allow an in-trim dive, while the dashed line shows a typical upper ocean thermal profile during a diving operation between depths A and B.

tion A) with no change in ballast. The ship sinks rapidly, losing buoyancy as it passes through the isothermal region due to hull compression. As it moves through the thermocline, added buoyancy allows the submarine to come into trim as it finishes its dive at location B. This is an example of a very rapid and efficient dive that makes full use of the ambient thermal conditions. An inexperienced submarine captain may dive less efficiently, discharging excessive amounts of ballast in the isothermal surface layer, where the pressure effect is greater than the buoyancy effect, and flooding again below the thermocline, where buoyancy exceeds pressure effect. It is obvious that efficient submarine diving operations, in addition to ASW considerations, depend on a knowledge of the ocean thermal structure.

The examples of the application of physical oceanography to military uses of the ocean that we have outlined are only a few of the many now being used by naval forces throughout the world. In the United States a major portion of the nation's marine research, both applied and basic, is funded by the Navy. Since tactical and strategic naval decisions are influenced to a large degree by environmental conditions, the growth of a nation's ocean science activities has been linked to the Navy's need to understand its operational environment, the ocean. Hence, military oceanography will continue to be an important area of applied oceanography in years to come.

CONCLUDING REMARKS

During the past century the descriptive and theoretical bases for physical oceanography have been firmly established. As ongoing and new studies continue to compete for available funding, each will more frequently have to be justified in terms of the ecological, economic, and governmental policy issues it addresses. This change in emphasis will lead to an expansion of the emerging field of applied oceanography. Under this new direction physical oceanographers will play a key role in helping society keep pace with the accelerated global need for the benefits that can be derived from the sea.

REFERENCES

Adams, E. E., K. Stolzenbach, and D. Harleman. *Near and Far Field Analysis of Buoyant Surface Discharges into Large Bodies of Water*. Report No. TR-205, Parsons Laboratory for Water Resources and Hydrodynamics, Massachusetts Institute of Technology, Cambridge, 1975.

Auer, S. J. "Energy from the Sea." *The Gulf Stream*, **6–7**, National Oceanic and Atmospheric Administration, 3–6 (1980).

Barnett, T. P. "On the Generation, Dissipation, and Prediction of Ocean Wind Waves." *J. Geophys. Res.*, **73**, 513–529 (1968).

Barnett, T. P., and J. Wilkerson. "On the Generation of Ocean Wind Waves as Inferred from Airborne Radar Measurements of Fetch-Limited Spectra." *J. Marine Res.*, **25**, 292–328 (1967).

Bisagni, J. J. *Passage of Anticyclonic Eddies through Deepwater Dumpsite 106 during 1974 and 1975*. NOAA DER 76-1, National Oceanic and Atmospheric Administration, Rockville, 1976.

Bishop, J. M. *Formation, Solution, and Application of a Non-Linear Miles–Phillips Spectral Component Growth Model*. CGOU TR74-1, Coast Guard Oceanographic Unit, Washington, D.C., 1974.

Bishop, J. M. "The Utilization of Offshore Currents for Improved Ship Efficiency." *Mariners Weather Log*, **22-1**, National Oceanic and Atmospheric Administration, 9–12 (1978).

Bishop, J. M. *A Climatological Oil Spill Planning Guide. No. 1. The New York Bight*, National Oceanic and Atmospheric Administration; available from the National Technical Information Center #PB 80-157613, Washington, D.C., 1980a.

Bishop, J. M. "A Climatological Oil Trajectory Model." *Mariners Weather Log*, **24-5**, National Oceanic and Atmospheric Administration, 344–350 (1980b).

Bishop, J. M. "A Note on the Seasonal Transport on the Middle Atlantic Shelf." *J. Geophys. Res.*, **85**, 4933–4936 (1980c).

Bishop, J., and J. Overland, "Seasonal Drift on the Middle Atlantic Shelf." *Deep Sea Res.*, **24**, 161–169 (1977).

Bleick, W. E., and F. Faulkner. "A Survey of Numerical Ship Routing." In *Proceedings of the Naval Oceanographic Conference*, Monterey, 1971.

Breaker, L. "The Application of Satellite Remote Sensing to West Coast Fisheries." *Marine Tech. Soc. J.*, **15-3**, 32–40 (1981).

Bretschneider, C. L. "Revised Wave Forecasting Relationships." In *Proceedings of the 2nd Conference on Coastal Engineering*, ASCE, Council on Wave Research, Houston, 1952.

Bretschneider, C. L. *Revisions in Wave Forecasting*, Look Laboratory Report, University of Hawaii, Honolulu, 1970.

Bretschneider, C. L., and R. Reid. *Modification of Wave Height Due to Bottom Friction, Percolation, and Refraction*. Beach Erosion Board Technical Memo 45, U.S. Army Corps of Engineers, Washington, D.C., 1954.

Brucks, J. J., J. Butler, K. Faller, H. Holley, A. Kemmerer, T. Leming, K. Savastano, and T. Vanselous. *LANDSAT Menhaden and Thread Herring Resources Investigation*. SEFC No. 77-16F, MARMAP No. 145, NSTL Station, Mississippi, 1977.

Cardone, V. J. *Specifications of the Wind Field Distribution in the Marine Boundary Layer for Wave Forecasting*. TR-69-1, Dept. of Meteorology and Oceanography, New York University, New York, 1969.

Cardone, V. J., W. J. Pierson, and E. Ward. "Hindcasting the Directional Spectra of Hurricane Generated Waves." *J. Petrol. Technol.*, **28**, 385–394 (1976).

Clancy, R. M., P. Martin, S. Piacsek, and K. Pollak. *Test and Evaluation of an Operationally Capable Synoptic Upper Ocean Forecast System*. TN-92, Naval Ocean Research and Development Activity, NSTL Station, Mississippi, 1981.

Cornillon, P., M. Spaulding, and M. Reed. "Impact Assessment in Oil Spill Modeling." In *Proceedings of the Workshop on the Physical Behavior of Oil in the Marine Environment*, Dept. of Civil Engineering, Princeton University, Princeton, 1979.

Crutcher, H. L., and R. Quayle. *Mariners Worldwide Climate Guide to Tropical Storms at Sea*. NAVAIR 50-IC-61, Naval Weather Service and National Climatic Center, Asheville, North Carolina, 1974.

Csanady, G. T. "Mean Circulation in Shallow Seas." *J. Geophys. Res.*, **81**, 5389–5399 (1976).

Dennis, R. E., and E. Long. A User's Guide to a Computer Program for Harmonic Analysis of Data at Tidal Frequencies, NOAA TR NOS 41, U.S. Department of Commerce, National Ocean Survey, Washington, D.C., 1971.

Dickson, R. R., and H. Lamb. "A Review of Recent Hydrometeorological Events in the North Atlantic Sector." In *Proceedings of the ICNAF Environmental Symposium*, Bedford Institute, Nova Scotia, 1971.

Dobson, R. S. *Some Applications of a Digital Computer to Hydraulic Engineering Problems*. TR-80, Dept. of Civil Engineering, Stanford University, Palo Alto, 1967.

Earle, M. D. *Oceanographic Design Conditions for Offshore Light Towers*, TN 6110-8-75, U.S. Naval Oceanographic Office, NSTL Station, Mississippi, 1975.

EDIS Magazine, "The Eurocean ODA Project." National Oceanic and Atmospheric Administration, **11-4**, Washington, D.C., 1980.

Ekman, V. W. "On the Influence of the Earth's Rotation on Ocean Currents." *Arkiv. Mat. Astron. Phys.*, **2**, No. 11, 1–53 (1905).

Ewing, J. A., T. Weare, and B. Worthington. "A Hindcast Study of Extreme Wave Conditions in the North Sea." *J. Geophys. Res.*, **84**, 5739–5747 (1979).

Fay, J. "The Spread of Oil Slicks on a Calm Sea." In *Oil on the Sea*, D. Hoult, ed., Plenum Press, New York, 1969.

Galt, J. A., and G. Torgrimson. "An On-Scene Spill Model for Pollutant Trajectory Simulations." In *Proceedings of the Workshop on the Physical Behavior of Oil in the Marine Environment*, Dept. of Civil Engineering, Princeton University, Princeton, 1979.

Glantz, M. H., and J. D. Thompson, eds. *Resource Management and Environmental Uncertainty: Lessons from Coastal Upwelling Fisheries*, John Wiley & Sons, New York, 1981.

Goldberg, E. D., ed. *Proceedings of a Workshop on Assimilative Capacity of U.S. Coastal Waters for Pollutants,* National Oceanic and Atmospheric Administration, Boulder, 1979.

Gunther, H., W. Rosenthal, T. Weare, B. Worthington, K. Hasselmann, and J. Ewing. "A Hybrid Parametrical Wave Prediction Model." *J. Geophys. Res.,* **84,** 5727–5738 (1979).

Harris, D. L. *Characteristics of Hurricane Storm Surge.* TP-48, U.S. Weather Bureau, Washington, D.C., 1963.

Hasselmann, K., D. Ross, P. Muller, and W. Sell. "A Parametric Wave Prediction Model." *J. Phys. Oceanogr.,* **6,** 200–228 (1976).

Hess, W. N., ed. *The Amoco Cadiz Oil Spill,* National Oceanic and Atmospheric Administration and Environmental Protection Agency, Washington, D.C., 1978.

Holcombe, R. M., G. MacDougall, and I. Perlroth. "Appraisal of Minimum Time Tracks Based on Wave Conditions." *Mariners Weather Log,* **3-1,** National Oceanic and Atmospheric Administration, 1–4 (1959).

Inoue, T. *On the Growth of the Spectrum of a Wind Generated Sea According to a Modified Miles–Phillips Mechanism and its Application to Wave Forecasting.* TR-67-5, Dept. of Meteorology and Oceanography, New York University, New York, 1967.

James, R. W. *Applications of Wave Forecasts to Marine Navigation.* SP 1, U.S. Naval Oceanographic Office, NSTL Station, Mississippi, 1957.

James, R. W. *Ocean Thermal Structure Forecasting.* SP 105, U.S. Naval Oceanographic Office, NSTL Station, Mississippi, 1966a.

James, R. W. "The Hazard of Giant Waves." *Mariners Weather Log,* **10-4,** National Oceanic and Atmospheric Administration, 115–118, 1966b.

James, R. W. "Dangerous Waves Along the North Wall of the Gulf Stream." *Mariners Weather Log,* **18,** National Oceanic and Atmospheric Administration, 363–366 (1974).

Jelesnianski, C. P. *SPLASH—Special Program to List Amplitudes of Surges from Hurricanes,* National Oceanic and Atmospheric Administration, Rockville, 1972.

Johnson, J. H. "Food Production From the Oceans." In *Proceedings of the NMFS/EDS Workshop on Climate and Fisheries,* National Oceanic and Atmospheric Administration, Washington, D.C., 1976.

Kamenkovich, V. *Fundamentals of Ocean Dynamics,* Elsevier Scientific Publishing Company, New York, 1977.

Katz, E. J. "Effect of the Propagation of Internal Waves in Underwater Sound Transmission." *J. Acoustic Soc. Amer.,* **42,** 83–87 (1967).

Kemmerer, A. J., and J. Butler. "Finding Fish with Satellites." *Marine Fisheries Rev.* **39-1,** Government Printing Officce, Washington, D.C., 16–21 (1977).

Kinsman, B. *Wind Waves,* Prentice-Hall, Englewood Cliffs, 1965.

Kopenski, R. P., and E. Long. *An Environmental Assessment of Northern Puget Sound and the Strait of Juan De Fuca,* National Oceanic and Atmospheric Administration, Seattle, 1981.

Larson, S. D., D. Dale, and T. Laevastu. *Large-scale Turnover in the Oceans.* TR-1-71, U.S. Naval Environmental Prediction Research Facility, Monterey, 1971.

Laurs, R. M., and R. Lynn. "Oceanography and Albacore Tuna, *Thunnus alalunga,* in the Northeast Pacific During 1974." In *The Environment of the United States Living Marine Resources—1974,* MARMAP No. 104, Washington, D.C., 1976.

Leming, T. D. *Application of a Coastal Current Model to the Recruitment of Shrimp in the Northwestern Gulf of Mexico.* SEFC, NSTL Station, Mississippi, 1982.

Library of Congress, *Energy from the Oceans,* U.S. Government Printing Office No. 95-455, Washington, D.C., 1978.

Lissaman, P. B., R. Radkey, W. Mouton, and D. Thompson. *Evaluation of Hydroelastic and Dynamic Behavior of Key Components of the Ocean Turbine System.* Report to the U.S. Department of Energy, Washington, D.C., 1979.

Long, R. B. "Forecasting Hurricane Waves." *Mariners Weather Log,* **23-1,** National Oceanic and Atmospheric Administration, 1–10 (1979).

Maul, G. A. "Variability in the Gulf Stream System." *The Gulf Stream,* **2–10,** National Oceanic and Atmospheric Administration, 6–7 (1976).

McCormick, M. E. *Ocean Engineering Wave Mechanics. Ocean Engineering: A Wiley Series,* M. McCormick, ed., John Wiley & Sons, New York, 1973.

Miles, J. W. "On the Generation of Surface Waves by Shear Flows." *J. Fluid Mech.,* **3,** 185–204 (1957).

Nagle, F. W. *A Numerical Study in Optimum Track Ship Routing Climatology.* TP-10-72, U.S. Naval Environmental Prediction Research Facility, Monterey, 1972.

National Oceanic and Atmospheric Administration. *Ocean Dumping in the New York Bight.* TR-ERL-321-MESA-2, Washington, D.C., 1975.

National Oceanic and Atmospheric Administration. *An Analysis of Brine Disposal in the Gulf of Mexico.* Reports to the Federal Energy Administration, Washington, D.C., 1977a.

National Oceanic and Atmospheric Administration. *Long Island Beach Pollution: June 1976.* MESA Special Report, Washington, D.C., 1977b.

National Oceanic and Atmospheric Administration. *Storm Surge and Hurricane Safety.* U.S. Government Printing Office No. 003-018-00092-0, Washington, D.C., 1979.

National Oceanic and Atmospheric Administration. *Annual NEMP Report on the Health of the Northeast Coastal Waters of the United States, 1980.* TM-NMFS-F/NEC-10, Woods Hole, 1981a.

National Oceanic and Atmospheric Administration. *Assessment Report on the Effects of Waste Dumping in 106-Mile Ocean Waste Disposal Site.* SP-81-1, Rockville, 1981b.

National Oceanic and Atmospheric Administration. *Deep Seabed Mining, Draft Programmatic Environmental Impact Statement,* Washington, D.C., 1981c.

National Oceanic and Atmospheric Administration. *Final Environmental Impact Statement for Commercial Ocean Thermal Energy Conversion (OTEC) Licensing.* U.S. Government Printing Office, Washington, D.C., 1981d.

National Oceanic and Atmospheric Administration. *Deep Seabed Mining, Marine Environmental Research Plan 1981–85,* Washington, D.C., 1982.

NAVPERS. *Aerographer's Mate 1 and 2, U.S. Navy Training Course 10362,* Memphis, 1965.

Nelson, W. R., M. Ingham, and W. Schaaf. "Larval Transport and Year Class Strength of Atlantic Menhaden, *Brevoortia tyrannus.*" *Fish. Bull. U.S.* **75-1,** 23–41 (1977).

Okubo, A. *A Review of Theoretical Models of Turbulent Diffusion in the Sea.* Report TR 30, Chesapeake Bay Institute, The Johns Hopkins University, Baltimore, 1962.

Overland, J. "Providing Winds for Wave Models." In *Ocean Wave Climate,* M. Earle and A. Malahoff, eds., Plenum Press, New York, 1979.

Ozturgut, E., G. Anderson, R. Burns, J. Lavelle, and S. Swift. *Deep Ocean Mining of Manganese in the North Pacific: Pre-mining Environmental Conditions and Anticipated Mining Effects.* TM ERL MESA-33, National Oceanic and Atmospheric Administration, Boulder, 1978.

Pequegnat, W. E., L. H. Pequegnat, B. James, E. Kennedy, R. Fay, and A. Fredericks. *Procedural Guide for Designation Surveys of Ocean Dredged Material Disposal Sites.* EL-81-1, U.S. Army Corps of Engineers, Washington, D.C., 1981.

Phillips, O. M. "On the Generation of Waves by Turbulent Wind." *J. Fluid Mech.,* **2,** 417–445 (1957).

Pierson, W., and L. Moskowitz. "A Proposed Spectral Form for Fully Developed Seas Based on the Similarity Theory of S. A. Kitaigorodskii." *J. Geophys. Res.,* **69,** 5181–5190 (1964).

Pierson, W. J., G. Neumann, and R. W. James. *Practical Methods for Observing and Forecast-*

ing Ocean Waves by Means of Spectra and Statistics. H.O. 603, U.S. Naval Oceanographic Office, NSTL Station, Mississippi, 1955.

Pierson, W. J., L. Tick, and L. Baer. "Computer Based Procedures for Preparing Global Wave Forecasts and Wind Field Analyses Capable of Using Wave Data Obtained by a Spacecraft." In *Sixth Naval Hydrodynamics Symposium,* Office of Naval Research, Washington, D.C., 1966.

Prandtl, L. *Essentials of Fluid Dynamics,* Hafner Press, New York, 1952.

Richards, F., ed. *Coastal Upwelling,* Coastal and Estuarine Sciences, American Geophysical Union, Washington, D.C., 1981.

Richardson, W. S. *Forecasting Extratropical Storm-Related Beach Erosion Along the U.S. East Coast.* TDL-78-13, National Oceanic and Atmospheric Administration, Washington, D.C., 1978.

Ross, D. B. "A Simplified Model for Forecasting Hurricane Generated Waves" (Abstract). *Bull. Am. Meteorol. Soc.,* 113 (1976).

Schukraft, D. "Time Savings Through Oceanrouting." *Oceanroutes Newsletter,* Oceanroutes Inc., Palo Alto, 1978.

Schureman, P. *Manual of Harmonic Analysis and Prediction of Tides,* U.S. Department of Commerce SP-98, U.S. Government Printing Office, Washington, D.C., 1958.

Shonting, D. H. *Observations of Short Term Heating of the Surface Layer of the Ocean.* TM 308, U.S. Naval Underwater Ordnance Station, Newport, 1964,

Slack, J. R., T. Wyant, and K. Lanfear. *An Oil Spill Risk Analysis for the Southern California Outer Continental Shelf Lease Area.* U.S. Geological Survey 78–80, Reston, 1978.

Smith, R. A., J. Slack, T. Wyant, and K. Lanfear. *The Oil Spill Risk Model of the U.S. Geological Survey,* U.S. Geological Survey 80-687, Reston, 1980.

Sverdrup, H.U., and W. Munk. *Wind Sea and Swell: Theory of Relations for Forecasting.* H.O. 601, U.S. Naval Oceanographic Office, NSTL Station, Mississippi, 1947.

U.S. Army Coastal Engineering Research Center. *Shore Protection Manual,* Fort Belvoir, Virginia, 1973.

U.S. Coast Guard. *National Search and Rescue Manual.* CG-308, U.S. Government Printing Office, Washington, D.C., 1973.

U.S. Department of Commerce National Ocean Survey. *Tide Table for the Pacific Coast of the United States,* Washington, D.C., 1970.

U.S. Environmental Protection Agency. *Ocean Dumping.* 1979 Annual Report to Congress, Washington, D.C., 1980.

U.S. Naval Oceanographic Office. *Oceanography and Underwater Sound for Naval Applications.* SP 84, Washington, D.C., 1965.

U.S. Naval Weather Service Command. *Navy Marine Climatic Atlas of the World, Vols. I–VII,* Washington, D.C., 1965–1978.

U.S. Navy. *Oceanography for the Navy Meteorologist,* Naval Weather Research Facility, Norfolk, 1960.

U.S. Navy. *Outline of Synoptic Numerical Oceanographic Analysis and Forecasting Program at the U.S. Naval Fleet Numerical Weather Facility.* TM 5, Monterey, 1965.

U.S. Navy. *Pilot Chart of the North Atlantic Ocean (March),* Naval Oceanographic Office, NSTL Station, Mississippi, 1969.

U.S. Navy. *Tactical Applications Guide.* NEPRF TR 77-04, Monterey, 1981.

Von Arx, W. S., H. Stewart, and J. Apel. "The Florida Current as a Potential Source of Useable Energy." In *Proceedings of the MacArthur Workshop on the Feasibility of Extracting Useable Energy from the Florida Current,* H. B. Stewart, ed., Palm Beach Shores, Florida, 1974.

Weare, T. J., and B. Worthington. "A Numerical Model Hindcast of Severe Wave Conditions for the North Sea." In *Turbulent Fluxes Through the Sea Surface; Wave Dynamics and Prediction,* A. Favre and K. Hasselmann, eds., Plenum Press, New York, 1978.

Williams, J. *Introduction to Marine Pollution Control; Ocean Engineering: A Wiley Series,* M. McCormick, ed., John Wiley & Sons, New York, 1979.

Wilson, B. W. "Deep Water Waves Generation by Moving Wind Systems." *J. Waterways,* **WW2,** 113–141 (1961).

Wilson, B. W. "Numerical Prediction of Ocean Waves in the North Atlantic for December, 1959." *Deut. Hydrogr.,* **18,** 114–130 (1965).

INDEX